Kerosene Heaters

A Consumer's Review

by Peggy Glenn

Aames-Allen
PUBLISHING CO.
Huntington Beach CA 92648

KEROSENE HEATERS: A Consumer's Review
by Peggy Glenn

Aames-Allen Publishing Co.
924 Main Street
Huntington Beach CA 92648 USA
714/536-4926

Library of Congress Cataloging in Publication Data
Glenn, Peggy
    Kerosene Heaters.

    1. Kerosene Heaters.   I. Title.

TH7450.5.G54   1984   697'.24   84-276
ISBN   0-936930-56-X
ISBN   0-936930-57-8 (pbk.)

Table of Contents

2225732

Dedication

To my family . . .

    most especially Gary . . .

who provides all the warmth I'll ever need.

1

Introduction - Why This Book?

"Kerosene heaters -- are they safe?"

Commissioner Stuart Statler of the Consumer Product Safety Commission asked that question during an all-day commission hearing July 5, 1983, in Washington, D.C. The Commission's research staff answered with a 2-inch thick briefing package, a parade of experts, and a half day of testimony; the kerosene heater industry answered with four more hours of testimony and an equally impressive packet of paper and list of experts. The fire service presented written testimony later.

Have you made up your mind?

The press has been on a see-saw since mid-1982. One day you might read a newspaper article that would frighten you; a week later you might read a magazine article that would reassure you; at some point you might have seen a television commercial that presented the "new generation" portable kerosene heaters as the most sensible alternative to out-of-control fuel bills.

I have spent almost a year looking at all sides of the kerosene heater picture. Are they safe? Are they economical? Are they a health hazard? Are they more of a problem than any other type of alternative heating?

I'm a writer. My profession -- my job -- is to gather facts from every possible source and to present the information in an orderly and logical manner and in language that you can understand.

I'm also a convert to safety. In 1975 I married a fireman. At that time I had never heard of a smoke detector, had never conducted a home fire drill with my children, and wouldn't really have known a fire hazard if it had hit me -- and a few almost did.

Along the way I gradually became aware of errors in my own behavior -- errors of **comission** (doing the wrong things), or errors of **omission** (not doing the right things) that had the potential for endangering my own life and health and the lives and health of people around me.

I looked at myself one day and fantasized that I was "average," that probably 75% of the American people were just like me. I say like me, because they weren't safety-conscious like my husband. Folks like me went through day after day never giving a thought to safety.

Mind you, I didn't heed very many of my husband's directions or warnings on command. Every time he told me "don't do that!" I wanted to know why. Every time he told me "you must do it this way!" I again wanted to know why. Fortunately, he loved me enough to put up with all of my challenges. He patiently explained all of the whys. Slowly with an understanding of the "WHY," I began to consciously make the choice for safety.

In 1982, after seven years of attitude adjustment, my husband and I wrote a book on all aspects of family fire safety entitled, **DON'T GET BURNED! A Family Fire-Safety Guide.** Folks liked what we did, they enjoyed our common-sense approach and common-folk language. We traveled the country spreading the word about fire and burn safety through prevention. And we began to hear about the controversy over kerosene heaters. We had covered the subject only briefly in our big book.

That's why I wrote this book -- to fill that gap. I wrote this one alone, though. This is information from a journalist's perspective, not from the fire service perspective.

I am convinced that we're a nation of people who must know "why" and "what happens if." We're in the "Information Age." We want to have all the information and then make our own choices based on the information we have received.

John Naisbitt, in his best-selling book **Megatrends,** sees a new breed of people in the 80s. He sees individualism at its highest peak. He says that people have become disillusioned with their institutions (like me with my fireman rule-maker) and that these same people -- armed with enough information -- are relying more and more on themselves and other people to fill their needs.

As consumers, as rule-followers, we want to be treated as equals in the decision-making and behavior process. We don't want to feel as if we've been fooled or manipulated.

What I've tried to do with this book is give you all the information I've been able to find out about the "new generation" of portable kerosene heaters. I've talked with members of the fire service -- some for, some violently against, and some neutral. I've talked with many members of the Consumer Product Safety Commission's staff and with two of the Commissioners. I've spent a lot of time talking with heater manufacturers or distributors. I've talked with people who own and operate heater sales and service outlets. I've also talked with many consumers who have kerosene heaters in their homes.

I've tried to present everything as objectively and as fairly as possible. I've drawn conclusions, but you don't have to accept them. I've also asked questions throughout that will help you decide whether or not you'll buy and/or use a kerosene heater; and if so, whether or not you'll choose to use it safely.

In the final analysis, YOU -- the consumer -- are the biggest factor in the kerosene heater safety picture. A heater is only as safe as its user. This is true of almost any product: a car, a lawn-mower, a toaster, a frying pan. You must use a kerosene heater according to the manufacturer's instructions, you must fuel it properly and with the right fuel. Most of the heater design changes that have been recommended and will be imple-mented are a deliberate attempt to protect you from your own unsafe actions.

I hope you enjoy this book; I hope you learn something from it; I hope it helps you. This book is not a replacement for the manufacturer's oper-ating manual. This isn't a technical book. This is a tool to help you evaluate whether or not you'll buy a kerosene heater. It's a supplement to, and further explanation of, the safety precautions.

*   *   *

Are kerosene heaters safe? Read on and make up your own mind.

2

In Perspective:
The Controversy
The Heater's History

The modern, portable, kerosene heater . . .

In recent years, no appliance or device has been the subject of quite as much controversy or study.

Conservative sources estimate that approximately **9 million** modern, portable, "new generation" kerosene heaters were being used in U.S. homes by the end of the 1982-1983 winter heating season.

THE CONTROVERSY

In late 1982 and early 1983, newspapers and magazines, some using skull and crossbone illustrations, took up the kerosene heater issue. Using statements and information from the fire service, the Consumer Product Safety Commission, and kerosene-heater industry sources -- the three key figures in the controversy -- the messages were often conflicting. While some articles reported fire deaths, and issued somber warnings about fire, contact burn, and air pollution risks, other articles reported that there were no serious health or safety risks associated with kerosene heaters.

5

In the middle of the controversy was the consumer who was struggling to pay astronomical home heating bills. Traditional central-heating methods and traditional fuels -- electricity, fuel oil, and natural gas -- were rapidly becoming out of the reach of many budgets. Zone heating -- heating part of the living quarters part of the time -- became the most practical alternative for many people.

Where did all of this controversy leave the consumer? How was he or she to sort the fear from the fact and make an intelligent decision about whether or not to purchase and use a kerosene heater? How was he or she to combat soaring winter fuel bills? Was he or she going to defy local regulations in order to stay warm on a limited budget?

At all levels of government -- town, city, county, state, and federal -- lively debate accompanied any discussion or attempt to change existing regulations governing the sale and/or use of kerosene heaters in residential spaces.

A REVIEW OF THE THREE MAJOR POSITIONS
The Consumer Product Safety Commission
The Consumer Product Safety Commission (CPSC) mobilized a special task force to study the problem. They were concerned about the well-being of the person who bought, used, or was exposed to a kerosene heater. The CPSC's study centered on the fire, contact burn, and harmful emission characteristics of kerosene heaters. The CPSC gathered data from the national fire reporting system, from news sources, from its field offices, from an information network of hospital emergency rooms, and from consumers.

In July, 1983, the staff of the Consumer Product Safety Commission presented the findings of its intensive 18-month study to the Commission.

Their research had looked extensively at three specific areas: fires involving the heaters, cases where people had suffered burns by coming in direct contact with the hot heater cabinets, and the emission characteristics of the heaters.

To quote directly from their testimony and written report to the commission, "It is the staff's judgement that the most important element of kerosene-heater safety is the pattern of consumer use."

After exhaustive testing and research, the CPSC staff made the following specific recommendations to the Commissioners regarding the heaters:

A requirement for emergency shut-off devices, preferably ones which would operate independently of the wick adjustment mechanism and which would be durable and not subject to failure;

Design changes to eliminate the possibility of consumer error or intervention with regard to the manufacturer's recommended wick setting; specifically, making it impossible for the consumer to lower the wick below the level at which it would burn efficiently, cleanly, and safely;

A testing requirement that would set a limit on the allowable level of nitrogen-dioxide (NO_2) to reduce the potential for undesirable health effects;

Additional safety measures, such as guards or grills, to protect against both the incidence of, and the severity of, contact burns, particularly in young children;

A comprehensive consumer information and consumer education program to

explain kerosene-heater safety and to
address all the issues that had been
studied or considered during the staff
investigation.

The CPSC staff also recommended: upgrading
existing voluntary safety and emission and labeling
requirements; adoption of a voluntary standard for
containers that would be used by the consumer to
transport and store kerosene, and more widespread
distribution of the necessary 1-K kerosene fuel.

The CPSC "Bottom Line" -- as I interpret both
their written material and my in-person discussions
and research with them: as they stated in the
fourth sentence of their study's summary, the
heaters pass the tests with some recommendations.
The consumers, on the other hand, will make the
difference with regard to safety. The heaters are
safe as long as the consumers use them in a safe
manner.

Remember, that's **my** interpretation of their
written and verbal material.

The Fire Service -- National and Local
The fire service has always been involved in issues
which they feel might be life-threatening or
property-threatening to "their people." The fire
service is traditionally protective -- and almost
custodial -- of the people whom they feel are
within their safeguarding jurisdiction, the people
who trust them for safety and security.

Fire service involvement in the current kerosene-
heater controversy began in 1979 when the CPSC
was asked by the Newark, New Jersey, Fire
Department to ban portable kerosene heaters. The
CPSC turned down that request on a 4 to 3 vote,
essentially saying that there was limited information on the hazards, an apparent limited use of

ID:31833040899038
FICTION BENED
Copy:1
Almost / Elizabeth Be
\Benedict, Elizabeth.
due:11/10/2001,23:59

ID:31833039354755
FICTION MYSTERY EVANO
Copy:3
Seven-up / by Janet E
\Evanovich, Janet.
due:11/10/2001,23:59

ID:31833040816289
FICTION MYSTERY BLOCK
Copy:1
Hope to die : a Matth
\Block, Lawrence.
due:11/10/2001,23:59

ID:31833032033133
CD J VAS IGS N 88
Copy:2
In the garden of soul
\Vas (Musical group)
due:11/10/2001,23:59

ID:31833001235743
697.24 G48K
Copy:2
Kerosene heaters : a
\Glenn, Peggy.
 ue:11/10/2001,23:59

ALLEN COUNTY PUBLIC LIBRARY

TELEPHONE RENEWAL
(219) 421-1240
DURING MAIN LIBRARY BUSINESS
HOURS ONLY. HAVE YOUR LIBRARY
CARD AND MATERIAL READY

ALLEN COUNTY PUBLIC LIBRARY

TELEPHONE RENEWAL
(219) 421-1240
DURING MAIN LIBRARY BUSINESS
HOURS ONLY. HAVE YOUR LIBRARY
CARD AND MATERIAL READY

ALLEN COUNTY PUBLIC LIBRARY

TELEPHONE RENEWAL
(219) 421-1240
DURING MAIN LIBRARY BUSINESS
HOURS ONLY. HAVE YOUR LIBRARY
CARD AND MATERIAL READY

ALLEN COUNTY PUBLIC LIBRARY

TELEPHONE RENEWAL
(219) 421-1240
DURING MAIN LIBRARY BUSINESS
HOURS ONLY. HAVE YOUR LIBRARY
CARD AND MATERIAL READY

the product, and a new voluntary standard for product safety. The CPSC felt that mandatory regulation was unnecessary and inappropriate at that time. Their later look at the whole home heating safety issue, and kerosene heaters in particular, has already been discussed.

Fire officials weren't happy with the Commission's ruling; they expressed fear that the consumer was at great risk. They cited fire, burn injury, and property-loss hazards from heater use and from fuel storage. Many of their fears were based on overwhelming death, injury, and property damage statistics involving old-style kerosene heaters (those produced prior to the mid-1970s). In early 1983, the fire service mobilized a national reporting system to try and record and study all incidents resulting in contact burns, deaths, fires, or property damage involving kerosene heaters.

The fire service study consisted of collecting information from a limited number of cooperating agencies. Four national agencies, the International Association of Fire Chiefs (IAFC), the National Fire Protection Association (NFPA), the International Society of Fire Service Instructors (ISFSI), and the Fire Marshals Association of North America (FMANA) asked for voluntary reporting of all incidents that involved alternative heating equipment during the 1982-1983 heating season (September 1982 through April 1983).

The results of that random sampling were reported in the September 1983 issue of Fire Journal. In the study, alternative heating equipment was divided into two categories: **"solid-fuel" heaters:** fireplaces, fireplace inserts, wood stoves, and coal stoves; and **other heaters:** kerosene heaters, electric heaters, installed natural gas heaters, installed LP gas heaters, portable fuel oil heaters, and waste oil heaters.

To summarize the report of that random sampling, 740 fires were reported involving all types of alternative heating equipment: all solid fuel heaters - 453; kerosene heaters - 194; electric heaters - 59; fixed natural gas heaters - 14; fixed LP gas heaters - 13; portable fuel oil heaters - 4; waste oil heaters - 3. Civilian deaths from those 740 fires totaled 116. The report emphasized that the number of deaths per fire and per type of heating equipment should not be used to make safety comparisons or assumptions about the type of equipment. However, in my conversations with some particularly outspoken members of the fire service, those comparisons are indeed being drawn and cited as even more reason to ban kerosene heaters from consumer use.

Especially noteworthy in that report was data that broke down the alternative-heating-caused fire deaths by region, and by the breakdown of the cause of the fire in the case of kerosene heaters.

By region, the South, with slightly more than one-fourth of the reported fires, accounted for over one-half of the fire deaths. This lopsided share of deaths rings true with other facts that are known about fire deaths in the "Burn Belt," traditionally rural and low-income areas in the South where uneven climate is a big factor. Winters are not traditionally harsh in the South, and when cold weather comes, people throw caution to the wind and do whatever they can to keep warm -- often with disastrous results.

By cause of fire, ALL the reported incidents indicated consumer error. Sixteen fires from using the wrong fuel, twelve fires involved improper refueling technique, and four fires involved placing the heater too close to combustibles (things that will burn). Those facts support my earlier statement. People cause fires -- not products.

The Fire Service "Bottom Line" -- on the record, opposed. But in a verbal survey of nearly 100 fire people at all levels, although no one agreed to let me use his or her name, I found that:

> Fifty percent feel that the heaters are safe and that it's nearly impossible (or at least a monumental task) to educate and inform consumers so that they use them safely. Many of the people in this group would like to be more involved in consumer education, but in states where the heaters are illegal for residential use, the management of their fire departments forbid such action on the grounds that to educate about an illegal device is to sanction the use of that device.

> Twenty percent feel that the heaters are safe and that consumers can be educated to an acceptable level of safety with the heaters.

> Ten percent feel strongly that the heaters should be banned in all residential use.

> Twenty percent still hadn't made up their minds. They're waiting to see what happens next heating season.

Among the fire people most likely to say that the heaters are safe and that people can be taught to use them safely, many own kerosene heaters themselves and use them in their homes. Many also have woodstoves and/or small electric heaters.

Just as with the sample of fire incidents used by the fire service in its study, my sample is random, and may not represent true percentages of the feelings of the fire service as a whole. A random sampling is just that, random.

The Kerosene Heater Industry

In response to all the government charges, fire service charges, and press reports, the kerosene heater industry -- composed of manufacturers and importers/distributors -- also conducted an intensive study of the safety issues. At the conclusion of their 18-month, $800,000 study, they confirmed their initial belief that the products were safe, but that the fires, deaths, injuries, and damage that did occur were caused by consumers using the heaters in a variety of unsafe ways.

The industry's organization, the National Kerosene Heater Association (NKHA), also commissioned an exhaustive study of the kerosene heater's effects on indoor air quality. An October 1982 article in Consumer Reports had been very critical of the potential for pollution from kerosene heaters.

The results of that study, presented later in a little more detail with more emission information, essentially showed no severe health danger from using kerosene heaters, provided they were used according to manufacturer's directions.

The NKHA and its member companies were also very active in drafting voluntary performance and safety standards -- standards more strict than any that had been applied by a government agency or an independent testing laboratory. NKHA established its own heater certification program. Many of their standards were higher than the standards imposed by Underwriter's Laboratory (UL). NKHA also worked diligently with fuel suppliers to try and guarantee a reliable source of uncontaminated, low-sulfur 1-K kerosene fuel for heater owners.

Once all the studies had reported that the heaters were basically safe if used safely, the NKHA started a massive public relations campaign in an effort to correct the record and change the press and consumer opinion. They tried to overcome the bad publicity, restore the faith of heater owners, and point up the economic benefits of zone heating with kerosene heaters.

To work on the "people problem," the NKHA also began an ambitious consumer education program using public service announcements on television and radio, printed materials for use in retail stores, and extensive dealer support and training. Their message to the consumer: a reminder to be warm **and safe** in winter by carefully following all of the instructions packed with industry-approved kerosene heaters. NKHA also printed safety tip sheets for people who might have misplaced or misunderstood the original instructions.

The National Safety Council recognized these consumer education efforts and awarded NKHA and its members an honorable mention for its efforts in 1983.

Obviously, the NKHA had a reason for all of this. Between the bad press, a warm winter, and a drop in oil prices, their industry wasn't in the best of health. In addition, they had lost a lot of consumer trust. But most of all, it appears that they were firmly convinced all along that the heaters were safe and that the consumer deserved an alternative -- their alternative -- to high fuel costs from traditional heating methods or central heating systems. They spent a lot of money and time when they could just as easily have folded up in bankruptcy and left every retailer and consumer in limbo.

They could also have been uncooperative with the governmental agencies and dragged the issue through the courts for years. CPSC Chairman Nancy Harvey Steorts, was impressed with their cooperative attitude and actions. CPSC Vice Chairman Terence Scanlon also has spoken in favor of cooperation rather than regulation. The consumer is clearly the winner in all of this. Minor design changes for safety, massive education campaigns for safety, continued testing for emission safety, and ongoing cooperation with the CPSC seem to assure that the heaters will be even safer and that consumers will use them in the right way -- safety first, last, and always.

The NKHA "Bottom Line" -- as I interpret it, the NKHA continues to believe that the heaters are safe. They may also be more convinced than before that the key to acceptance and safe usage of the heaters is in an education program that will permanently impress upon the consumer the importance of proper selection by size and type of heater, safe operation of the heater, and diligent maintenance.

That mind-switch may be difficult in some cases. Some people will always be safety conscious, some people can be converted to safety, many people have an "it will never happen to me" attitude, and a large number of people have been spoiled -- by an abundance of "instant" products -- into acting without needing to think.

KEROSENE HEATERS: Are They Safe?

So far, the vote is a qualified "yes." Yes, the heaters are safe **IF** consumers use them safely.

In the rest of this book I'll tell you as much about the heaters as I know and I'll tell you all the whys for all the safety precautions. There aren't many, but they are very important.

* * *

A LITTLE HISTORY: Why Kerosene Heaters, Where Did They Come From? How Are The New Models Different?

The modern, "new generation," portable kerosene heaters were almost unknown in the U.S. until the late 1970s. According to conservative estimates by industry sources, almost **9 million** heaters were in use in U.S. homes by the end of 1983. Manufacturers and importers expect to sell another 3 million heaters before the 1984 heating season begins. Is all this the result of a clever advertising phenomenon? Where did they come from? Why do they sell so well?

The big push for alternative heating methods of all types can probably be traced to the Arab oil embargo in the early 1970s. The cost of heating homes by any method -- oil, electricity, or natural gas -- caused our now famous "energy crisis." First we tried conservation; no patriotic American had his or her home thermostat set above 65° in winter nor below 78° in summer. Then we tried insulation; we would protect against cold in winter and against heat in summer. But energy prices continued to climb in spite of the fact that we used less and less fuel.

Even though the number of households increased as "baby boomers" moved into their own homes, home energy use still dropped dramatically from 1978 to 1980. Americans have cut home energy consumption by 17%, but are paying 27% higher bills. Clearly, the answer was to switch to non-traditional methods and alternative fuels.

Americans had been spoiled by seemingly endless supplies of energy at reasonable prices. First the OPEC crisis and then government de-regulation of utilities caused the price of home heating energy to soar. However, in other parts of the world, particularly Japan and Western Europe, consumers were far more advanced in the technology they used to heat their homes. Japan, for example, has virtually no native fuel source. No strangers to new technology, the Japanese developed the kerosene heater as we know it now. The heaters they use are tested to a very demanding and comprehensive list of safety standards. Today more than 75% of Japanese households use kerosene heaters as their primary heating method.

Kerosene heaters are also used extensively in Western Europe as a supplemental home heating method. In Europe, in contrast to America, a gallon of gasoline costs almost 3 times as much. The same is true for nearly all petroleum products.

Many people credit the meteoric rise in U.S. use of kerosene heaters to Bill Litwin, owner and founder of the now-bankrupt Kero-Sun company. Litwin was a commercial airline pilot who spent a great deal of time traveling to the Far East and Europe which is where he became familiar with the wide acceptance of the heaters in those areas. He formed Kero-Sun to import and sell kerosene heaters in the U.S. Kero-Sun spent a lot of money on advertising, and was hit hard when the heaters started to receive bad publicity.

By 1982, there were more than fifty companies selling kerosene heaters in the U.S. However, close to 70% of the market was shared by just eight companies; more than forty-five companies split the remaining 30% of heater sales. As of this writing, some of the smaller companies which were underfinanced or unable to spend large amounts for advertising, have gone out of business. Others have been bought out by larger firms.

Some of the companies which survived the combination of bad publicity and a warm winter were the ones which had other products besides kerosene heaters. Some firms which saw the negative trends early were able to sit tight and wait it out.

What this all means for the consumer, is that the companies that survive through 1984 can probably be relied on to provide superior products and reliable service. Remember, that while the CPSC made recommendations for future heater modifications, it did not ban nor issue recall orders for any of the heaters already in use. If you already own a heater, or if you buy one before the latest safety modifications are added, your heater dealer and manufacturer are important links in the safety chain. Before you buy a "close out" heater, be sure you'll be able to have it serviced and find replacement parts for it.

Because America is the home of capitalism and entrepreneurship at its best, many small firms have climbed aboard the "quick-money express." Therefore, one other very big concern is that you avoid buying any imported products that do not display safety labels or certification labels. More on that in Chapter 3.

A WORD ABOUT OLD LEGISLATION
One more small aspect in the controversy is the ban in some areas on home use of all unvented heaters, including old-style kerosene heaters, and the application of that ban to the present-day kerosene heater. The design of the new heaters has changed considerably. The new models bear almost no resemblance to the old models except that they all use kerosene fuel and most models use a wick.

However, the bans that are in effect are virtually unenforceable short of violating your constitutional right against unlawful search and seizure of your home. In addition, no local or state agency has the staff to go door-to-door to confiscate illegal heaters. The laws that have been proposed by the NKHA are an attempt to protect the consumer by allowing only safety-approved models to be imported or sold.

If the subject comes up in your state or your local area, be an informed voter. Let your legislator or local politician know how you feel. By buying a heater, 9 million people have made a choice.

*   *   *   *

Before we close this chapter, stop for a moment and consider one other element in the kerosene-heater "controversy."

After the OPEC oil embargo and U.S. government deregulation (which shot oil, gas, and electricity rates to record levels in the late 1970s), and after several consecutive cold winters, consumers bought wood stoves, fireplace inserts, electric heaters, and kerosene heaters in record numbers in order to stay warm and avoid personal bankruptcy.

1. Why did consumer's use alternative heaters?
    A: To cut down on winter heating fuel bills.

2. How many consumers had switched to "alternative" heating methods?
    A: Approximately 25 million.

3. If consumers weren't spending money for natural gas, electricity, or heating oil, who lost?
    A: The utility companies and oil companies.

Because of earlier bans on "old style" kerosene heaters, which of the alternative heating methods seemed to be the easiest "target," and who seemed the most likely "gunslingers"? You decide!

How to Choose a Kerosene Heater

Who will use it?
How much heat do you need?
What type of heater to buy?
Where to buy a heater?

The modern, portable kerosene heater. If you listen to the advertisements, these heaters are:

easy to use

inexpensive to operate

much safer than older models

not messy, and don't require clean-up

the answer to your high winter heating bills

ALL OF THE ABOVE

True. Kerosene heaters have a place in the line-up of supplemental and emergency heating equipment. BUT, you the consumer, should be informed about the different kinds of heaters, and aware of your own heating needs and personality traits in order to make an intelligent buying decision. Your satisfaction with a heater lies in (1) your ability to choose the right heater for your use; (2) your attention to safety; and (3) your willingness to follow all the manufacturer's directions.

EVALUATING YOU AS A HEATER USER

Several factors influence the size and style of the kerosene heater you'll buy. Before you even look at your heating needs, however, take a look at you. Are you a person who pays attention to details and instructions, are you safety-minded, are you maintenance conscious? Or are you like I used to be: "don't make me read about it, just let me plug it in and use it."

Kerosene heaters may be about 20 years late in making their American debut. Since the early 1960s, we have been bombarded with products that we're told will require less thought and give us more free time. Everything is "instant." A kerosene heater is part instant and part nostalgia. But nostalgia requires a little change in our attitude. We need to be slow and careful. "Instant" is not always completely compatible with "nostalgia."

When we sit in front of a fireplace or stoke our wood stove or use a kerosene heater we should never forget the awesome power of heat and open flames. Our ancestors grew up respecting heat and fire, and they made safety part of their daily routine.

A kerosene heater is not complicated, you don't need an engineering degree to use it. But it does require that you pay strict attention to the manufacturer's instructions and that you never override the safety precautions or take short cuts. If it says "wait 15 minutes before refueling," that's what it means, no matter how cold it is, no matter how hurried you are. If it says "do not lower the wick below the recommended level," it means don't lower the wick for any reason. If it says, "use only good quality 1-K kerosene fuel," it means just that, no other fuel will work. But you have to believe it and follow the rules.

A kerosene heater must also be maintained carefully according to the manufacturer's instructions. The heaters must be cleaned before being stored for summer, the wick must be changed, and the heaters must be checked again before being put into service in the next heating season. Again, you don't need to be a fix-it genius, but familiarity with a screw driver, pliers, and a few simple tools is helpful.

If you stiffen with fright at the thought of tinkering with anything mechanical, that doesn't mean you should never buy or use a kerosene heater. Instead, purchase your heater from an authorized dealer and refer all of your care and maintenance questions and tasks to that dealer. Paying the dealer for labor and parts once or twice a year will add a little bit to your cost of operating a heater, but it's money well spent if it keeps you from making a dangerous mistake.

EVALUATING YOUR HEATING NEEDS

Now that you've decided you're the type of person who would give a kerosene heater the care and respect it deserves, you're ready to figure out what type of heater you need and how large it needs to be.

Heater output is rated in BTUs, or British Thermal Units. Before making any style decision about your heater, you should first determine how much heat you will need it to produce. Generally you can come close to figuring out which heater will deliver the necessary amount of BTUs simply by knowing how many square feet of space you wish to heat and multiplying that number by 28. For example, if your living room is 12 feet long by 15 feet wide, 12' x 15' equals 180 square feet.

Then, multiplied by 28, you need a heater that will deliver close to 5,000 BTUs.

Notice that I said "generally" and "close to." There is another very important factor to consider when selecting the size of the heater you'll buy. What is your personal "comfort range" -- the air or room temperature at which you're the most comfortable? To reach your "comfort range" inside your home during the winter, are you willing to keep your home a little cooler and wear a sweater? Will crawling children be warm enough on the floor? If not, buy a heater with more BTU output.

This example may explain "comfort range." My husband says that I'm one of those people with a "faulty thermostat." In winter, when almost everyone else around me is warm, I'm reaching for a sweater or a lap robe. And in summer, when others are dripping with sweat, I'm comfortable.

My "comfort range" is somewhere between 75 and 81 degrees F. My husband's "comfort range" is closer to 70 degrees F. What is yours? How does your ideal comfort temperature match up with the rest of the people in your family?

Using your own or your family's average "comfort range" as a gauge, choose a heater that delivers just under or just over the BTU amount your calculations say you'll need. Remember also that except for some new models, most kerosene heaters are not thermostatically controlled. They are "ON" or they are "OFF." The thermostat is you.

In addition, a wise shopper will consider a few more things as well. Look carefully at the building type, construction, and insulation; the location of the room or rooms you'll be heating, and the activity in those rooms; and consider all of the following factors and their reasons:

- Is it a stone or wooden building?

    Stone construction absorbs heat like a sponge. Select a heater with a slightly higher BTU rating than you would need for a wooden building.

- Is the building well or poorly insulated?.

    A well-insulated building holds heat from a small heater. A drafty building requires a larger heater.

- Is the room on the first or second floor?

    Heat rises. First-floor space will require a larger heater than second-floor space.

- Does the room have common or outside walls? How much help will you get from the sun?

    Outside walls are heat escape routes and bear the brunt of winter winds. Rooms with inside walls stay warm longer. Morning sun will help warm a room.

- The traffic flow in and out of the area.

    Each person who opens and shuts an outside door creates a heat escape. Lots of traffic and heat escape requires a larger unit to keep the area warm.

- How much furniture is in the room?

    Furniture absorbs heat. An empty room will heat up more quickly, but a full room will stay warm longer.

- How much activity goes on in the room?

    A family sitting still watching television will need more heat than a family playing pool in the all-purpose room.

Make a chart listing all of these plus or minus factors to help you decide the size of heater you need.

HEATER TYPES

There are two major types of kerosene heaters on the market: (1) the **unvented,** or portable, heater; and (2) the **vented,** or stationary, heater. This book relates to the portable, unvented heaters which make up the overwhelming majority of heaters in use today. Concerns about fueling, emissions, and supervised use are common to the unvented heaters. The end of this chapter contains a short description of vented heaters and their proper use and installation.

Another set of terms frequently used to describe kerosene heaters are: (1) the **convection** heater, and (2) the **radiant** heater. The principles of radiant heat and convective heat overlap just slightly simply because of the laws of physics. Radiant heaters create a small amount of convective heat simply because as the objects and people in the room become warmed by the radiant heat waves, the air temperature becomes warmer, too, creating air movement. And all convection heaters produce a small amount of radiant heat. The difference is in the proportion of heat that is radiant or convective.

Radiant heat travels quickly and heats objects that are directly in front of the heat source. When you place your hand in front of a radiant heat source, the side facing the heat will warm quickly while the other side of your hand remains cool.

A convection heater actually heats air. The warm air moves up, forcing cooler air down. This circular motion of the air eventually warms all the air in the room. The warm air also transfers heat to the objects and people in the room.

The sun is a good example of radiant heat. On a cloudy day, you can sit on the lawn and be comfortable in the open because the temperature of the air is "just right." If the sun peeks through the clouds, you will feel the extra heat, even though the air temperature may stay the same.

The greatest advantage of convective heat rather than radiant heat is the absence of hot areas in a room. Generally convective heat creates the same sensation as central heat. On the other hand, the convective heat process takes a long time to warm up a room and all its people and objects to the desired comfort range. Radiant heaters can deliver on-the-spot heat to a specific area nearly instantly.

The Convection Heater
As the term implies, a convection heater is omni-directional, and the heat is moved in all directions away from the heater. The heat that is produced creates air flow which continually spreads warm air away from the heater in all directions while taking in cooler air from all directions.

Convection heaters are best suited for large open areas which require a lot of heat and which have enough room for people and furniture. Because convection heaters spread heat in all directions, you are not able to place one of these models near a wall nor closer than 3 feet in any direction to furniture, draperies, fabrics, people, or anything else that will burn. Convection heaters are usually easy to spot by their round shape.

All convection heaters that I have seen have a built-in fuel tank. The entire heater must be moved in order to refill the fuel tank. Remember this fact as you go through a selection "checklist." Are you strong enough to lift the heater and move it smoothly without tipping it from inside the house to an outdoor fueling area and back inside

Convection heater

Photo courtesy of Alladin Energy Products, Inc.

This particular radiant heater will deliver enough BTUs to heat approximately 300 square feet. Using my comfort range example, and supposing I lived in a drafty home where the outside temperature was often below freezing with a strong wind battering three outside walls with no morning sun, I suspect that this heater's efficiency would be just about right for a room of 200 square feet.

Photo courtesy of Glo International.

to operate it? If not, you may wish to purchase a set of dolly wheels to move the heater. One caution about wheels, however. As of this writing, no heater on wheels has received a UL listing or NKHA approval. The heater's built-in bottom-heavy design is a safeguard against accidental tip over. Placing that heater on wheels and leaving it there as you operate the heater may seriously alter that safety feature.

The Radiant Heater

A radiant heater moves the heat very rapidly in one direction away from the heater. Most radiant heaters use a polished reflector which is located behind the burner assembly. The reflector is usually curved and is placed the whole length of the "back" wall of the heater. As the burner assembly creates heat, the warmth "bounces" off the reflector and is forced into the room.

Because these heaters produce heat to the front, rather than on all sides, a radiant heater can be placed closer to a wall than a convection heater. Radiant heat travels a long distance quickly. Objects that will burn should be kept at least 3 feet away from the front of a radiant heater. A radiant heater is best suited for rooms with heavy traffic flow and small spaces where a convection heater would not be safe in the middle of the room.

Radiant heaters are usually rectangular and look like a box or cabinet. In addition, most radiant heaters have removable fuel tanks; however, some radiant heaters are similar to convection heaters in that they have permanent fuel tanks. If you would be unable to safely move the heater because of its weight, you may opt for a radiant heater that has a removable fuel tank. For information on the safe way to fill either a convection heater or a radiant heater, refer to pages 50-54.

SELECTING FOR SAFETY FEATURES, QUALITY ASSURANCE

The next step in deciding which heater to buy is looking for some very important safety features. Some of these safety features are:

- An automatic shut-off device which is triggered whenever the heater is tipped over or jarred.

  > In some heaters the shut off device may lower the wick, in other models the shut off device actually severs the wick mechanism from the burner assembly. This "guillotine" method is sometimes more reliable than the "drop wick" mechanism. The drop wick mechanism can get hung up if the wick is crusted with carbon deposits. If the model you select uses the drop wick method, be certain that you keep the wick clean at all times. With either shut-off mechanism, check the spring lever every time you light the heater.

- Guards or grills to protect you from touching the hot surfaces of the heater and being burned.

- A low center of gravity to prevent the heater from being accidentally knocked over.

  > Remember as mentioned earlier, that the addition of wheels may interfere with this important safety feature. Use wheels only when moving the heater.

- A tamper-proof wick adjustment mechanism to assure that the wick burns cleanly, evenly, and is not subject to sudden flare-up.

  Remember that you can't lower the heat output of the heater by lowering the wick. Turning the wick below the manufacturer's recommended position sets up fire and pollutant problems.

- Look for a heater with sturdy metal construction.

  A kerosene heater goes through extreme temperature changes. On a cold morning, the heater may be as cool as 40°. After the heater has been running a short time, the cabinet temperature may reach 400° or more. A well-built heater is designed to withstand these temperature changes without warping.

IN WICKLESS HEATERS ONLY:
- An oxygen depletion sensor (ODS) which shuts down the heater if the oxygen level in the room becomes dangerously low. Some larger wickless heaters are equipped with sensors which turn off the heater if the temperature in the heater gets too high.

  To avoid problems of insufficient oxygen or high levels of pollutants, always operate the heater in an area that has adequate indoor or outdoor ventilation.

You may also purchase additional fence-like cages to keep small children from crawling too close.

Also check to see that the heater you buy has a label showing that it has been certified or listed by a nationally-recognized testing laboratory. As discussed in the first chapter, both the Underwriter's Laboratory (UL) and the National Kerosene Heater Association (NKHA) have recently adopted rigid heater guidelines. Appearance of both certifications on a heater tells you that the heater has passed the strictest tests that were in effect at the time the heater was manufactured.

These listing labels also tell you that the manufacturer's instruction and care manual is factual and easy to understand.

Both UL and NKHA are evaluating their current guidelines whenever they have new information. If you already have an kerosene heater that does not have these certification labels, you may wish to write to the manufacturer or UL or NKHA to see if your heater meets current specifications for safety. Addresses are in the last chapter, "Resources."

When kerosene heaters first began to re-appear in the late 1970s, UL's standard 647 had not been adopted and NKHA had not begun its certification program. These heaters may display the OSHA label and a label from JHAIA, the Japanese Heating Appliance Inspection Association which were the only standards in force in the late 1970s. A heater may also bear certification from additional private testing laboratories such as PFS (Product Fabrication Service) of Madison, Wisconsin.

ABOVE: Examples of U.L. listing labels that may be on your heater.

BELOW LEFT: An example of the Japanese Heating Appliance Inspection Association's certification label.

BELOW RIGHT: An example of one label from PFS (Product Fabrication Service).

WHERE TO BUY A HEATER

Many of the larger manufacturers or importers of today's kerosene heaters sell only through authorized dealers. These manufacturers feel very strongly that a kerosene heater's safety record depends on the way the consumer uses and maintains the heater. They feel that selling through conscientious dealers is the way to insure that you are adequately informed of all the safety measures and supported with reliable service.

If you buy your heater from a specialty store, the chance is high that the store owner or employees will help you through each step of the selection process, instruct you thoroughly in all aspects of safe and efficient operation of the heater, and provide annual maintenance for your heater. The heater dealer has a direct source for replacement parts, and can usually help you resolve any difficulties very quickly. The heater dealer is also very likely to have an influence on the fuel supply in your city or town as well.

You may see kerosene heaters in discount stores, home and hardware chain stores, and catalogs. This doesn't automatically mean that the heater is inferior, but it may well mean that you are on your own for operation, service, and maintenance of that heater once you've purchased it.

Most heater stores will service your heater even if you didn't buy it from them, although you might expect that their regular customers will occasionally be given preferential treatment. Of course, if you move from one area to another, be sure to ask your current heater service technician for a recommendation in your new location. If this isn't possible, write to the manufacturer and ask for an authorized dealer in your new area. Ask new neighbors for recommendations, too.

A vented kerosene heater is nearly always permanently installed as in the drawing below. This permanence allows for good air circulation inside the home or building, but also means that once the heater is installed, it stays. Vented kerosene heaters are equipped with fans which pull in cold, outside air to be used in the combustion process, heated and then released into the room. The gases from the combustion process are then forced back outside by fans. The outside venting eliminates the need to provide open-door ventilation inside the building.

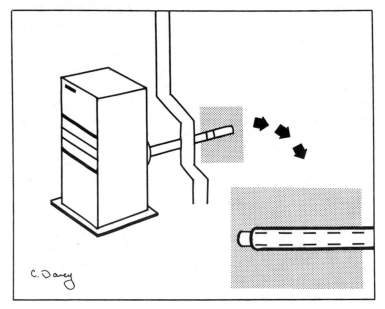

When installing a vented kerosene heater, you must be sure to comply with all local building codes. You may wish to have your heater installed by a professional heating appliance contractor who will obtain the necessary permits. This is no job for an amateur. If you have the skill to do the job, however, be sure to obtain a building permit and have all phases of the work inspected to be sure it is done safely.

4

How Do Kerosene Heaters Work?

Fuel system
Burner assembly
Ignition system
Shut-off device

As I've said before, the designs of the new portable kerosene heaters have changed greatly from the models that were popular in the 1940s. Old style kerosene heaters were usually round with bottom fuel tanks and very simple burners similar in design to kerosene lamps. This once-simple bottom-tank convection heater has evolved into a number of more complicated versions, though. The new versions have been completely re-designed for maximum safety and efficiency. However, some of the improvements, such as removable parts and complex burner assemblies, have opened the door for consumer error.

If you understand a little about how they work and why you must follow the directions carefully, you will probably not encounter difficulty with your heater. If you have a good idea of the mechanism that creates the heat and the fuel that runs the mechanism, you'll respect it, you'll take care of it, and you won't be a statistic -- not a headline-making statistic. You'll be one of those "millions of satisfied customers."

THE WICK

The lower portion of the wick (the skirt) rests in the fuel tank or sump. Kerosene from the tank is absorbed by the wick and carried up the wick by capillary action. This process takes 25-35 minutes in most models. The top of the wick is usually made of fiberglass strands. Fiberglass is used at the top of most wicks because it will withstand a flame, but will not be consumed by the flame. During the burning process, the fuel that reaches the top of the wick is heated by the flame and vaporizes. The vapors then burn (more about fuel vaporization in the next chapter.)

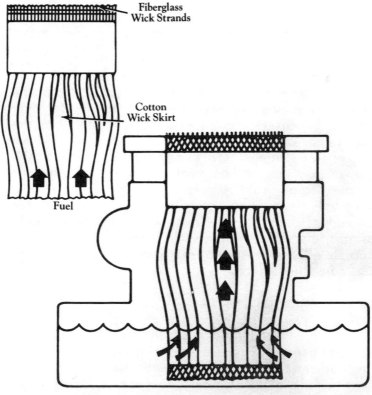

Fiberglass
Wick Strands

Cotton
Wick Skirt

Fuel

TOP: The wick by itself showing fuel direction.
BOTTOM: The wick as it relates to the burner
              assembly.

Precise adjustment of the wick height is critical to the safe and efficient operation of a heater. A wick that is **too low** will not allow the kerosene to vaporize correctly. This causes "low burning," a situation of uneven heating in the burner assembly which causes the metal parts to warp out of shape. This distortion then causes excess air flow which creates a dangerous "flare up" condition. The first sign of low burning is a strong odor. Also, low burning doesn't allow complete combustion of the fuel; this may cause pollution problems.

A wick that is **too high** will burn with a noticeable red flame, will produce smoke, and will cause unpleasant odors and fumes. This is a waste of fuel and may cause the wick to wear out very rapidly.

Check the wick condition often for carbon build-up and let the heater "burn dry" (see next page) about every five gallons worth. After each "burn dry" (fiberglass-wick heaters only), clean the wick gently with a brush as directed in the operation manual. Replace the wick at least once per heating season, more often if it becomes frayed.

The heater's operation manual has specific instructions about the proper wick height in the burner assembly (sometimes called a chimney). Wick heights vary from heater to heater, so check your manual closely. Your heater has an adjustment knob or mechanism for raising or lowering the wick. After you have set the wick at what appears to be the correct height, use a ruler to check the wick all around the edge. Measure at four or five different spots to be sure that the wick is sitting straight. Some of the latest models do allow more than one wick setting, depending on the level of heat you wish. However, most models have only one correct setting. Most heaters are turned off by lowering the wick so that it does not burn at all. Again, consult your owner's manual.

THE FUEL SYSTEM

As discussed in more detail elsewhere, your heater is equipped with either a built-in or a removable fuel tank. Either system provides a fuel supply which is below the wick. As the fuel is drawn up the wick during operation, the fuel level lowers in the tank. When the fuel flow stops, the wick "burns dry," and all burning stops. This "burn dry" process is recommended on a regular basis. Check the fuel level before each use. A tank of fuel will last between 10 and 20 hours depending on the size and design of your heater.

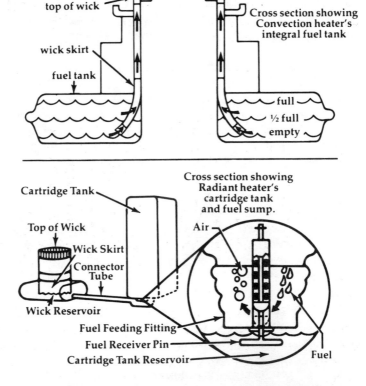

TOP: Fuel flow system in a convection heater with a built-in bottom tank.
BOTTOM: Fuel flow system in a radiant heater with a removable fuel tank.

Good quality fuel is absolutely vital to the proper operation of your heater. Always insist that the fuel you buy is high quality, water clear, uncontaminated 1-K kerosene. Nearly all trouble with a kerosene heater can be traced to either the fuel, the wick, (or the fuel's action on the wick), or the operator. As shown below, the kerosene is drawn up into the very tiny spaces at the fiberglass top of the wick. The spaces (and proper height adjustment) are what provide the ideal air-to-fuel mixture so that the kerosene burns completely. The diagram on the right shows how contaminated, poor quality, or dirty fuel can leave deposits on the wick. These deposits interfere with the capillary action and with the "perfect" burning environment.

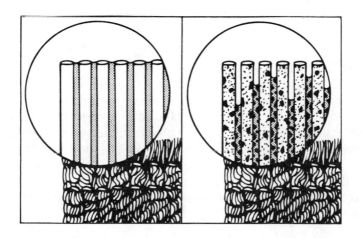

At the beginning of the heating season, you may encounter poor fuel that has been stored through the summer. Kerosene deteriorates with time and is also subject to bacteria growth. The fuel may become discolored, another reason for always giving it the "water-clear" test. Deteriorated or bacteria-infested fuel is difficult, and perhaps impossible, to light. If it does burn, it will give off excessive smoke and unpleasant odors.

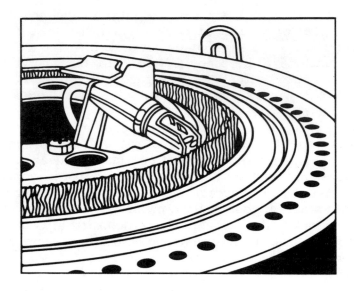

AUTOMATIC IGNITION SYSTEM

An automatic ignition system is a consumer con-
venience that is available on many models. It is
a very simple electro-mechanical system that is
powered with batteries or electric house current.
The ignition system is a safety feature so that
you don't have to use a match. The automatic
ignition system operates much like the electronic
pilot light on the new gas clothes dryers or pilot-
less gas stoves.

When the ignition switch or lever is pushed, the
plug is briefly turned on and gives off enough heat
to ignite the vaporizing kerosene. This then produces sufficient flame (and heat) to ignite the
remaining kerosene vapors around the entire circle
at the top of the wick, and the burning process
is maintained until the heater runs out of fuel or
the wick is lowered as discussed earlier. The
"glow plug," or igniter, should never touch the
wick.

THE BURNER ASSEMBLY

The burner assembly controls the air/fuel mixture -- the air flow that controls the rate and degree of combustion. This may be also be referred to as the "mantle," or the "heat chamber," or the "chimney." The inside of this section usually has several rows of round metal plates that have holes in them. The air moves through these holes. Occasionally these holes become plugged up with dust, animal hairs or other house debris. If your heater won't operate properly, you might also check this area to be sure it is clean.

The burner assembly on your heater may be glass enclosed, it may be open, or it may be metal with a mesh dome as pictured above.

AUTOMATIC SHUT-OFF SYSTEM

Heaters which have been approved for sale and use in America **must** be equipped with an automatic shut-off device. The device is designed to be activated any time the heater is tipped over, bumped, or jarred excessively. You may remember that the current generation of heaters were developed in earthquake-prone Japan. Any violent natural or man-made movement is dangerous.

The device may operate in one of two ways: (1) by dropping the wick below the burning position, or (2) by separating the wick and burner assembly from the fuel supply. This is an extremely important safety feature of your heater and you should check this system each time you use your heater. **Do not use the heater if the automatic shut-off system is broken.** Have it fixed immediately.

5

Kerosene Fuel - Only 1-K!

Buying the right fuel
Storing kerosene safely
Filling and refilling the heater
Fuel additives — yes or no?

In the very first chapter, I talked about the two most important factors in owning and operating a kerosene heater safely: the right fuel and a safety-conscious owner. Let's talk specifically now about the fuel.

WHAT KIND OF FUEL?

Today's modern kerosene heater is designed to be operated with low-sulfur, uncontaminated, water-clear, 1-K kerosene fuel. **NOTHING ELSE!**

If you are ever unsure of the quality of the fuel you purchase or if you have reason to suspect that the fuel has been contaminated or is left over from last year's stock, question the dealer where you bought it. He will probably, in turn, question his or her supplier. Your heater will not operate cleanly, efficiently, or safely without the right fuel. As the consumer, you must take the responsibility for knowing what you're buying, and you must not settle for less than what you need.

NEVER USE ANYTHING BUT 1-K KEROSENE FUEL IN YOUR HEATER.

Never use jet fuel, gasoline, camp-stove fuel, home heating oil, or any other liquid or solid fuel in your kerosene heater. By far, the greatest number of heater accidents were the result of using the wrong fuel, usually gasoline.

To help you understand why gasoline is so dangerous compared to kerosene, and why even trace amounts of gasoline in the kerosene fuel container can be disastrous, let's discuss some of the differences between gasoline and kerosene. Although they are both petroleum products, they are "skimmed off" at different points in the refining process and have unique chemical and burning characteristics.

Liquid fuels such as gasoline and kerosene do not burn in liquid form. When the liquids are heated to a specific temperature, the liquids give off vapors, and the vapors are what burn. This burning of the vapors then causes the temperature to rise which allows the remaining liquid to vaporize faster, creating a roaring explosive fireball.

The "flash point" is the lowest temperature at which, when the fuel starts to vaporize, those vapors will ignite ("flash") if there is an ignition source present. An ignition source may be: a lighted match, a cigarette, a pilot light, static electricity, or any other open flame or similar phenomenon.

The "flash point" or point at which gasoline will begin to vaporize, is -32°F. When gasoline is stored below -32°F, it will not vaporize, so it is relatively safe. BUT, how often is the temperature in your area consistently lower than 32 degrees BELOW zero! Almost never, so at any temperature above minus 32 degrees, gasoline will vaporize. Then, given the right mixture of oxygen, gasoline vapors, and ignition source, gasoline is a liquid bomb. Consider this fact: one gallon of gasoline is equal in explosive power to **6 sticks of dynamite**!

Kerosene, with a flash point of 120°F is easier to store and somewhat safer to use, but kerosene must never be diluted with another fuel, contaminated by another fuel, or allowed to be heated to a temperature that is above 100°F. To help put this idea into better perspective, look at the table below. The table shows the effect on the flash point whenever gasoline and kerosene are mixed.

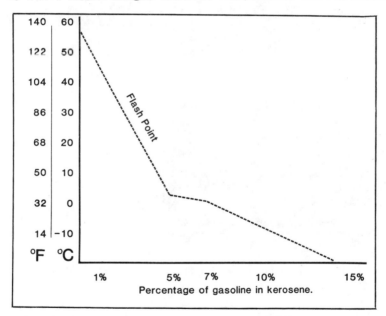

The flash point for gasoline is -32°. The flash point for kerosene is 120°F. With such a wide spread in the vaporization characteristic of the two fuels, the flash point for kerosene is lowered dangerously whenever even tiny amounts of gasoline are added to the fuel.
Source: Consumer Product Safety Commission

When purchasing bulk kerosene at a service station or hardware store, you may notice signs which resemble these. Always ask if the fuel is 1-K kerosene, and always look for a distinctive marking on the fuel pump.

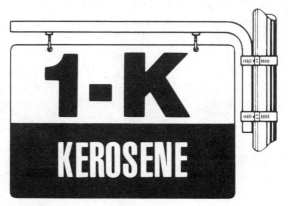

Graphics courtesy of Creative Products International, Tampa, Florida.

When kerosene heaters were first introduced in the U.S., 1-K kerosene was difficult to find. However, the right fuel is becoming more and more available as the demand for the fuel increases. You'd be wise to check the service stations, hardware stores, and other outlets in your area to be sure the right fuel is available before you purchase a heater.

Prepackaged kerosene, although a bit more expensive, is an alternative to buying bulk kerosene and having to carry the container back and forth. Depending on your safety habits and those personality traits we discussed earlier, this may be a more practical method for you.

Photo courtesy of Mark Quality Products, Lake Bluff, Illinois.

STORING THE FUEL

Always store kerosene in an approved container in a cool, dry, well-ventilated area that is outside your home. Always lock the storage area to keep out children, pets, and vandals.

Approved kerosene containers are usually blue and white, the same colors as the cover of this book. Only lined metal and high-density polyethylene cans are acceptable. Metal cans must be lined to protect against rust which will foul the fuel and cause operational problems with the heater. Some experts argue for a lined, rust-proof metal container, other experts prefer a poly container. At this writing, they are both acceptable and choice of a container is your personal preference. Photo courtesy of Keromate Products, Inc.

One important note to remember about fuel storage: You may store fuel during a heating season, but you should not stockpile fuel for several seasons nor store fuel through the summer. Kerosene deteriorates over time. Judge your fuel supply so that you'll run out of kerosene at about the same time as Spring finally arrives to stay. If you have kerosene left in your heater, burn the excess. If you find it absolutely necessary to drain your heater, use a siphon pump. Never dispose of kerosene into the drain or sewer system. Leftover kerosene fuel may safely be added to your home heating oil or to the fuel tank of **diesel** cars and trucks. If you don't have these options, call your local fire prevention officer.

HANDLING THE FUEL

Kerosene is a liquid fuel. By its nature, liquid fuel presents different problems from solid fuel (wood or coal) or gaseous fuel (natural or LP gas). To understand this, visualize two pitchers, one filled with water, one filled with ice cubes. If you spill the water from waist-high, it splashes over a wide area. If you spill the ice cubes from waist high, they will scatter over a wide area, also, but it's much easier to pick them up before they create a mess or cause any damage.

When using wood or coal as a fuel, if you spill a load as you walk in from the garage or back porch, you've maybe bruised a toe and maybe scattered wood splinters and scraps on the floor, or perhaps you have a black dusty mess to sweep up. If you spill kerosene, however, cleanup is more difficult and may present dangerous possibilities for fire.

For this reason, kerosene heaters should never be filled in the living quarters. One more problem with spilled kerosene is that as the fuel soaks into a carpet or rug, the wicking action lowers the flash point of the kerosene. The fuel is now spread over a wider area, it is diluted by absorption into the fabric, and will ignite more readily than in a liquid/vapor state.

FILLING OR REFILLING THE HEATER

Always fill and refill your heater outside the living space of your home, preferably out of doors. When the wind is swirling snow or sleet and the temperature hovers at zero, you'll be tempted to drag the fuel container into the house rather than dragging the heater or fuel tank out. This is unsafe. If you remember all of the reasons behind fueling the heater out of doors, you'll turn up the thermostat in your home and wait for a break in the storm. Think of this: If you have a major accident with the fuel, you'll be a lot more inconvenienced if your home burns down in the middle of a storm.

It's a good idea to get into the habit of refilling the heater before each use. By doing so, you are assured of several hours of uninterrupted warmth. As stated earlier and repeated later in the safety tips section, the heater should not be left on when no one is at home nor when you are sleeping.

Some kerosene heaters have removable fuel tanks. If your heater does not have a removable tank, then the entire unit must be moved outside in order to refill the tank safely. When moving your heater, be careful that you don't tilt or tip the heater. In some models, this could cause fuel to spill out of the tank and overfill the sump. In other models, the spilled kerosene may leak onto the floor or carpet or may become heated enough to flare up.

The safest and easiest way to fill or refuel your heater is with a siphon pump. You might be inclined to try to use a funnel and save the expense of a siphon pump, but safety experts don't advise it. Use of a funnel almost always results in spilled fuel which must be cleaned up, could stain the fill area, could leak out when the heater is moved back into the house, and could also ignite under the right conditions.

Some heaters are sold with a siphon pump; sometimes dealers include a pump with each heater they sell. If yours breaks or you need to buy a spare pump or a more sophisticated model, check the heater sales section of your hardware store, home center store, or heater sales store.

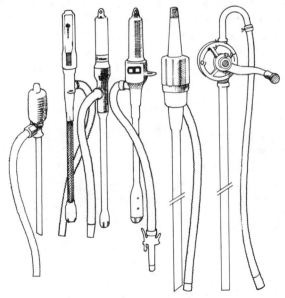

Typically, a siphon pump costs from about $4 for a small hand pump up to $60 for a heavy-duty pump. One battery operated model, which automatically shuts off to prevent tank overfills, retails for $30-$35.

Photo courtesy of Keromate Products, Inc.

**REMINDER! always fill the
heater's fuel tank OUTSIDE!**

Use only 1-K kerosene fuel.

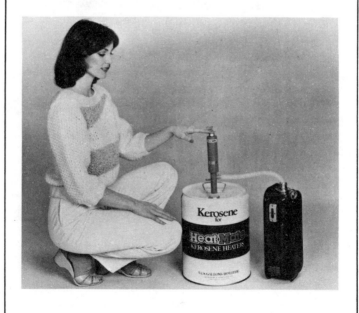

To use a siphon pump, insert the rigid end into
the fuel container, and insert the flexible end into
the heater's fuel tank.
Photo courtesy of Kupanoff & Associates, Inc.

When filling your heater, never fill past the "FULL" mark on the heater. As discussed earlier, as the fuel warms up, it will expand and give off more vapors. You must leave room in the fuel tank for this expansion. If you don't, the fuel could leak out around the tank creating a fire hazard and staining your floor or carpet.

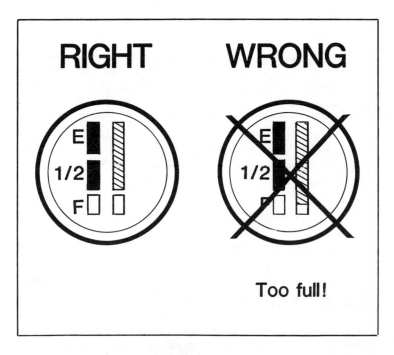

RIGHT **WRONG**

Too full!

Never attempt to refill your heater while it is operating. Always turn off the heater and allow it to cool before you refill the tank. This is true for models with built-in tanks and models with removable tanks.

If your heater runs out of fuel while in operation, wait until the heater is cool enough for you to touch. Although the room may cool down during this time, putting on an extra sweater is safer than filling a hot heater. As mentioned previously, if the weather outside is brutal, don't be tempted to bring the fuel inside. Instead, turn up the thermostat, put on extra clothes, and wait out the storm. Remember that your own body produces approximately 50 BTU's per hour. Find a way to capture that heat.

ADDITIVES

Many stores sell replacement parts, fuel siphon pumps, and fuel additives. Spare siphon pumps are always a good idea, the replacement wicks and parts may or may not fit and may or may not work (always check with your manufacturer); but the verdict isn't in yet on fuel additives. The commonest additives are those that contain alcohol, ether, nitrogen, organometallic compounds (heavy metals), or chlorinated hydrocarbons.

One of the greatest dangers of adding anything to the kerosene fuel, is that you'll change (and probably lower) its fuel point. For instance, while additives which contain either alcohol or ether may eliminate water contamination, these same chemicals will dangerously lower the flash point.

The remaining additives (nitrogen, organometallic compounds, and chlorinated hydrocarbons) are designed to slow down the formation of sludge and reduce the carbon build-up on the wick. However, these products were originally designed for use in engines rather than heating appliances. Engines do not ordinarily operate in confined indoor spaces, so the extra pollutants are not a health problem.

But today's kerosene heaters are designed to be used indoors. As stated earlier, no one agrees yet on indoor air quality standards. And all the tests that are being done on kerosene heaters involve the use of 1-K fuel in a heater that is operating at maximum efficiency. Until more studies have been done, you may be courting serious trouble by introducing even more products of combustion into your home's air.

Most heater dealers do not recommend the use of any fuel additive with the heater. Also remember that when the heaters were tested for their UL and NKHA listings, they burned 1-K kerosene fuel, no additives.

Scented additives are potentially hazardous, also. As mentioned in the previous chapter, a strong or unpleasant odor is one of the first signs of trouble in a kerosene heater. If this odor is masked with perfumed additives, your heater could be severely damaged (and you could be subjected to unhealthful levels of pollutants) if the heater continues to operate incorrectly. When the fuel is vaporizing correctly and burning at maximum efficiency, you will notice virtually no odor. There is a slight odor when you light it and when you turn it off as the vaporization process begins or ends.

The bottom line on additives: they are an added expense, they do not produce more heat, they may or may not work as intended, they may or may not be safe from a fire standpoint, an emission standpoint and a maintenance standpoint.

If you do choose to use an additive, be alert to any changes in the way your heater operates, and be alert for any unusual physical symptoms such as burning or itching eyes, breathing difficulties, headaches, nausea. Children, people who smoke, people with respiratory or heart problems, and elderly people are particularly susceptible. If you're in doubt about the use of fuel additives in your heater, check with the person who will offer service on your heater or write to the customer service center of your heater's manufacturer.

TROUBLE SHOOTING GUIDE

Possible Cause	Heater will not light	Flame goes out	Burner doesn't glow bright red	Smoke or odor	Flame height too low	Flame height too high	Heater will not go out	Heavy movement of wick adjuster knob	Remedy
Out of kerosene	•	•							Check fuel gauge, add kerosene
Wick adjusted too low	•	•	•		•				Adjust wick height w/wick adjuster knob
Water in kerosene	•	•	•	•				•	Drain tank, replace wick, refill with kerosene
Excessive deposits on wick	•	•	•	•				•	Clean wick, replace if necessary
New wick not saturated with kerosene	•	•	•	•					Allow additional soak time
Contaminated kerosene	•	•	•	•	•			•	Drain tank, replace wick, refill with ASTM 1-K kerosene
Burner not seated properly			•	•					Rotate burner to position properly
Ignition plug damage	•								Replace ignition plug
Batteries are dead	•								Replace batteries
Wick adjusted too high				•		•			Lower wick height with wick adjuster knob
Wick installed at incorrect height			•	•	•	•	•		Reinstall wick at proper height
Wick shaft damage							•	•	Replace wick shaft

6

Heater Use and Maintenance
Economy tips
Safety always
Regular care

After spending at least $100 dollars on your heater, your next thought is how to make the best use of that heater and how to insure that it will last as long as possible. In the financial world, this is called R-O-I, "Return on Investment." A more down-to-earth term would be **COMMON SENSE.**

Can a heater really save you money? Yes, if you choose the right size for the heating job, and if you make the best use of the heat that your model produces.

Typically, the cost-per-gallon of kerosene fuel is between $1.00 and $2.00 for bulk or at-the-pump service and around $4 if pre-packaged in disposable containers.

Depending on the size and BTU output of your heater, a gallon of fuel will provide 10-18 hours of heat. If we cut that right in the middle, a $1.50 gallon of fuel will last 14 hours, so your cost is 10.7¢ per hour. If you ran the heater 15 hours a day (7 am to 10 pm) the bill would be

$48.15 for a 30-day month. However, that's only for the kerosene fuel. If you use your central heating system to bring the whole house to a "survival" or maintenance temperature and use one or more kerosene heaters for additional "zone" heating, be sure to add the central heating system's operation and any other heater cost to your monthly heating bill.

Depending on the price you currently pay for electricity or natural gas, kerosene may or may not be less expensive. Use the following chart to help figure out if you'd really be saving money by using kerosene heaters, or only feeling better about the money you were spending because you weren't spending it all in one place.

Cost Comparison of Kerosene with Other Fuels

Kerosene per gallon 135,000 BTU/gallon	$1.00	$1.20	$1.40	$1.60	$1.80	$2.00
Electricity per kilowatt hour 3412 BTU/KwHr	2.6¢	3.1¢	3.6¢	4.1¢	4.6¢	5.2¢
Natural Gas per therm 100,000 BTU/therm	53¢	63.5¢	74¢	84.5¢	95¢	$1.06
Coal per ton 26,000,000 BTU/ton	$138	$165	$193	$220	$248	$275
Fuel oil per gallon 140,000 BTU/gallon	74¢	89¢	$1.04	$1.19	$1.33	$1.48

This cost comparison is based on the following fuel efficiency ratings:
kerosene 98% • electricity 100%
natural gas, coal & fuel oil 70%

To use the chart, first find out the local price per gallon for 1-K kerosene fuel. Then, check with your electric company and find out the price you are paying **per kilowatt hour** for electricity. Be sure to ask for the total of the basic rate plus any "cost adjustment." If you use natural gas, find out the cost per thermal unit (again the total of the basic rate plus any cost adjustments). If you use coal or home heating oil, again, find out what you pay per ton or per gallon. Once you know the per/gallon price for kerosene, look down the chart to either electricity, natural gas, coal, or fuel oil and compare that price to the price your utility quoted.

Using this chart, if kerosene costs you $1.40 per gallon, and your electric rate is 2.6¢ per kilowatt hour, the kerosene is more expensive. However, if you pay $1.40 for kerosene, and your electric utility charges more than 3.6¢ per kilowatt hour, kerosene is less expensive.

With central heating systems, energy experts estimate that you can cut your heating costs by 3% for every 1 degree that you lower your thermostat, no matter what fuel you use. So, if you lower the thermostat 10 degrees, from 75° to 65°, you save 30% on your central heating bill.

Now that you've lowered the thermostat and are relying on the kerosene heater to warm up the occupied sections of the house, there are just a few more tips to make the best use of the kerosene heater while assuring that the rest of the house doesn't suffer. If you have turned your thermostat down to a maintenance level, you may wish to relocate the thermostat to one of the rooms that is being "maintained." Why? Because if the thermostat is in the main living area, and that area is warmed to 75°F by the kerosene heater, the central heating system will not come on. The rest of the rooms in the house will never get any heat. They will not be "maintained" at all. Then, as these rooms get colder during the day and early evening because they haven't received any heat from the central heating unit, they will "rob" the living area of its heat. The kerosene heater will not be able to deliver enough BTUs to warm the living area, in spite of the fact that it is the "right" size for the room. Pretty soon no room in the house is warm enough and you'll feel as if the kerosene heater is not doing its job.

To solve this problem, install a thermostat in one of the rooms that is to be maintained at your chosen temperature. **Caution:** Be aware that all thermostats rely on a small amount of electric current to operate. Although the low voltage may not be enough to cause a deadly shock, you should consult a heating specialist or electrician for help in moving the thermostat to another location. A professional can probably do the job in a few hours. The small cost for hiring someone else to move the thermostat will be worth it if: (1) you avoid a costly accident; and (2) you start to save dramatically on your heating bills while keeping all the rooms comfortable.

If you have baseboard electric heaters, each with its own thermostat, this won't be a problem. The heaters in the "maintained" rooms will keep these less-used rooms at the maintenance level.

If you have high vaulted ceilings, heat from any source will collect at the highest point. You may be dollars ahead in both winter and summer by installing ceiling fans. Also, if your central heating unit has a "fan only" setting, you can put it to use by circulating the warm area from the living spaces to all rooms in the house. This will help with ventilation and air exchange, both of which are necessary to improve indoor air quality.

In deciding if a kerosene heater is your best bet economically, also consider the start-up one-time costs:

> the heater
>
> fuel container
>
> siphon pump (if not packed with heater)

And certain on-going expenses such as:

> trips to get fuel
>
> wick replacement
>
> annual maintenance

There are times and places where a kerosene heater is not practical, not safe, and not cheaper. For instance:

- In a home that is designed with no large, open common areas. Many older homes or "add-on" situations create boxy little rooms. Even a kerosene heater with the lowest BTU output will provide too much heat for most rooms in this type of home. Too much heat is a waste of your money.

- In a high-rise or multiple-family apartment building. With no outdoor place to store fuel and fill heaters, you place all the tenants or occupants in danger by storing fuel inside or by refilling your heater tank inside.

- In an area where the utility is less expensive and/or you must travel far to get kerosene, you haven't gained anything.

- In a home with several small children who cannot be effectively supervised to guard against contact burns or against accidentally tipping over the heater.

The trouble-shooting chart on page 56 will alert you to several maintenance routines that can help assure that your heater gives you good service for several years.

General Maintenance Tips

The flame on a heater almost always needs some fine tuning in order to assure complete combustion and clean, odorless burning. Make sure the wick is installed correctly and that the whole burner assembly is sitting straight. Then set the wick height precisely according to the direction in the operations manual. As stated earlier, check the wick condition frequently; if your wick is fiberglass, perform the burn-dry or "clean burning" procedure about once every 5 gallons.

If your heater uses a cotton wick, periodically trim the top of the heater with sharp scissors. Remove any uneven or brittle parts.

REMEMBER: **Never mix procedures on wicks.**
Cotton wicks should never burn dry.
Fiberglass wicks should never be trimmed.

Never substitute one wick type for another.
Replace a cotton wick with a cotton wick. Replace a worn fiberglass wick with another fiberglass wick.

High-grade 1-K fuel, free of impurities and low in sulfur, will add to the wick life, whether it's cotton or fiberglass. Nearly all success or failure, all joys or all problems, with kerosene heaters are traced to the wick, the fuel, or a combination of the two.

Keep the kerosene heater clean and rust-free. When the heater is turned off, gently vacuum the heater cabinet to remove hairs, dust, and debris which might hinder the air/fuel mixture.

7

Three Safety Issues
Fires
Contact Burns
Indoor Air Quality

Safety has been stressed in every chapter in this book. These next two pages will briefly review most of the concerns and prevention methods for fires, contact burns, and indoor air quality.

FIRE PREVENTION

William Overby of the U.S. Fire Administration has been quoted as saying, "It's not so much the way the devices are manufactured as the way people use them." Fire-safety tips include:

-- Never mix any other fuel with the kerosene.
-- Never fuel the heater while it is running or while it is still hot.
-- Never use any fuel but kerosene in a heater.
-- Always wipe up any spilled fuel.
-- Never place the heaters close to furniture, drapes, or anything else that will burn.
-- Always turn off the heater at bedtime or when no one is home.
-- Do not use the heater as a dryer for mittens, shoes, etc. The heater is designed only for heating.

-- Never use aerosol sprays or flammable liquids around a kerosene heater.
-- Do not allow trash or papers to accumulate near the heater.
-- Do not store solvents, paints, or any flammable products near a heater.
-- If you encounter any fire condition, get out of the house and call the fire department. Do not move the heater or attempt to put out the fire with water or blankets.

BURNS
-- Do not stand too close to a heater. The heat may feel good, but you can quickly become numbed to the heat's effect, just as a dip in the ocean quickly numbs the feet.

-- Children and older people are particularly vulnerable to exposure burns and burns from falling into or against a heater. If you feel this is a hazard in your home, erect a fence-like cage around your heater for extra protection.

EMISSIONS - INDOOR AIR QUALITY
Any appliance or device that relies on combustion (fire and flame) will produce gases and particles that will contaminate the air. The same heating device will use oxygen from the room during the burning process. There are really only two ways to provide adequate oxygen replacement and reduce your exposure to undesirable contaminants.

(1) Use only low-sulfur 1-K kerosene fuel and make sure the wick is clean and properly adjusted.

(2) **Always** maintain an adequate air flow in and out of the room where the kerosene heater is running. **Always** leave a door or window slightly ajar. **Never** operate a heater in a confined space, in a bedroom, or in an area without any ventilation. Warming a little fresh air costs only a few pennies.

8

General Safety - Part 1
Smoke detectors
Fire extinguishers

No matter what method you use to heat your home, you should be completely familiar with smoke detectors and fire extinguishers. The following questions and answers have been reprinted in shortened form from the information in <u>DON'T GET BURNED!,</u> the fire-safety book I wrote with my husband.

Officials estimated that smoke detectors were installed in 60% of American homes by 1982 -- a big improvement over a dismal 15 percent in 1978, but that means that 40% of American homes are still <u>not</u> protected. Most of the unprotected homes are lived in by folks who need the protection the most: the poor, the elderly, and the handicapped. In addition, even in homes that do have smoke detectors, many people don't know what the detector is, how to maintain it, how to respond to it, or how to buy another one if that one quits?

Smoke detectors definitely save lives. Early warning is the only way to survive a fire. Newspaper headlines and stories often point up the fact that a family did or did not have a smoke detector, and were or were not able to survive a fire. On the next several pages are questions and answers you may have asked asked about smoke detectors.

What is a smoke detector?

A smoke detector is basically an alarm that will buzz loudly when it detects smoke or any of the products of combustion (toxic gases or particles). The typical smoke detector is about as big as 3 or 4 slices of bread piled on top of one another.

First-Alert Smoke Detector. Photo courtesy BRK Electronics, a division of the Pittway Corporation.

Why should I have one?
How much does one cost?

To protect you and your family while you sleep. Most home fires occur between midnight and 6 a.m., and start in the living areas of the house or apartment, rather than in the bedrooms. The best way to protect you as you sleep is to monitor the air in your house and alert you before smoke or toxic gases get so thick that you can't survive. Most fire victims die of smoke inhalation, not burns.

A smoke detector is probably the greatest protective device that's available today. It won't prevent a fire; it won't protect your property if you're not home; and a detector won't put out a fire for you. What it **will** do is increase your chances of waking up, getting out safely, and calling the fire department for help. Cost: usually less than $20.

How do they work?

There are two basic types of detectors that are available to the public: those using an ion chamber, and those using a photoelectric principle.

Ion Chamber detectors detect products of combustion. They can detect a fire even if there is no visible smoke. The ion chamber detector uses very little electricity, and until recently was the only detector that could be effectively battery-operated. Photoelectric detectors detect visible particles of smoke.

Some experts argue in favor of the ion chamber detector, and others favor the photoelectric detector. Recently some manufacturers introduced models that use both principles in one detector and can be either battery-operated or hard-wired directly into the home or apartment current. Either detector will give you that extra margin of escape time.

How many should I have?
Where should I put it or them?

Ideally: one detector on each floor of your home (except the attic unless it's a finished attic or used as a bedroom). If you live in a single-story house that is spread out with more than one sleeping area, install a detector at the entrance to each sleeping area. If you have a high vaulted ceiling, put a detector near the top of that ceiling and one more in the hall leading to the bedrooms.

Is there anything special I should look for when buying a detector?

Yes. If your state requires it, make sure that the package has a notice saying that it is approved by the State Fire Marshal. Call your fire department if you're unsure about local regulations. The box should also contain a U.L. label. Instructions should be thorough and easy to understand and tell you how to install, test, and maintain it. If you buy a battery-operated model, check to see that it uses ordinary batteries.

Can I put it up myself?

If you've chosen a battery-operated model, you need a screwdriver and an hour to put it up. If you're unable to do it, ask a friend, or hire someone to do it for you. If you've chosen a model that is wired to the house current, hire an electrician to install it for you. Just as with relocating the thermostat, the last thing you want to do is get electrocuted.

Okay, it's up, now what? How much work is it to take care of?

Test it once a month on a regular basis. Choose the 1st of the month, the 15th, or any other convenient number. To test the detector use a small candle or a match. For the ion chamber detector, let the flame burn. For the photoelectric detector, light and then extinguish the candle or match and let the smoke filter into the detector. After the buzzer or alarm goes off, blow or fan the smoke away to silence the detector. This tests the chamber as well as the alarm battery.

Once a year, change the batteries and any other temporary parts. This can be any special day you choose, New Year's Day, your birthday, your anniversary, Ground Hog Day, whatever works. If your detector is battery-operated, make sure you always have a fresh set of batteries on hand. Many times, a fire has occurred when there were no batteries in the detector. Twice a year gently vacuum the detector's outside case so that if there's a fire, the smoke or gases can reach the detection unit inside the case.

Should I set if off for fire drills?

Yes, and no. It should be set off when you first get it and then at least once a year when all family members are home so that they'll recognize the sound when it goes off. But there's no need to set it off during family fire drills.

What if it breaks?

If your detector is so sensitive that it goes off from cigarette smoke or cooking odors, it needs to be moved to another location, repaired, or replaced. Don't just get angry and throw it away. It's made to save your life, not control it. If your detector breaks, replace it. Don't be without your detector while you wait for your original one to be repaired or replaced.

What do I do if it goes off?

Follow all the steps in your planned Home Escape Alive Drill (see the next chapter). Remember that your escape plan needs to be almost instinctive. When you've been wakened from a sound sleep by the loud buzzing of your detector, you're going to be a little groggy and confused. Knowing what to do well in advance of having to do it will make doing it a lot easier.

Right now skip to the USE YOUR H-E-A-D chapter. Write up your family's escape plan. Talk about it, diagram it, walk through it, practice it. You may feel a little foolish, and you may also feel a little squeamy talking about a fire in your home, but a little bit of time spent right now will pay off when you need it the most.

Who else should I tell?
Any time that you have an overnight guest, mention the smoke detector and explain the emergency exit plan from the room in which your guest will be sleeping. Also tell your neighbors about your detector -- for two reasons. First, they might hear the alarm when you're not home and call the fire department and possibly save you from a greater loss. Second, you may convince them to get one of their own if they don't already have one.

I'd like to buy one as a gift for someone else. Will they think I'm weird?
Possibly. If you're buying a detector for someone you care deeply about, they'll probably understand. If they won't accept your gift, though, donate it to a senior citizen center or give it as a gift at the next office party. Someone will appreciate it.

I can't afford one. What can I do?
First, smoke detectors aren't very expensive. Several fully-approved models sell for about $10. Second, many areas have programs to put free smoke detectors into high-risk, low-income homes. Last, many areas have local ordinances which REQUIRE landlords to install smoke detectors in EVERY living unit at no cost to the tenant. If your area has such a law and your landlord has not complied, notify the authorities immediately. You may be entitled to a detector under the law.

I travel a lot, is there anything for me?
Yes. At least 5 companies market travel smoke detectors. Cost: again, between $10-$15.

But won't my dog or cat save me?

It's possible, but unfortunately, more cats and dogs have been lost in fires than have saved the occupants. In 1980 more than 39,000 households lost pets in fires.

I'm deaf, how will I hear it?

There are experimental models available that have been quite reliable in waking deaf people. When the detector senses trouble, it goes off and also signals a receiver that turns on a warning strobe light, a vibrating device, or a light.

But I'm so careful. Fire won't happen to me.

That's what almost 6,000 people say every year. Fire very well may happen to you, or to someone you love. Fire doesn't choose its victims. It doesn't care if you're rich or poor, sick or well, famous or ordinary. Some of this country's most famous people have been victimized by fire: artist Norman Rockwell, former President Richard Nixon, Senator Howard Baker, actor Jack Cassiday, famous heart surgeon Dr. Michael Debakey, baseball star Dave Winfield, basketball legend Kareem Abdul Jabbar, and actor Gene Kelly.

EXTINGUISHERS

On the next few pages are the same kinds of questions and answers for fire extinguishers.

What is a fire extinguisher?

A pressurized cannister device that releases water or chemicals to put out a fire. They range in size from about the size of a beer can, to industrial-strength ones that are about 2 feet high and weigh 15-25 pounds.

What kind of fire can I fight with one?

Depending on the type of extinguisher you buy, you can fight either of the three classes of fires:

<u>Class A fires</u>
green label symbol   ordinary combustibles

<u>Class B fires</u>
red label symbol   flammable liquids

<u>Class C Fires</u>
blue label symbol   electrical

<u>Class D fires</u> are those involving combustible metals which are not usually found in your home. These fires require a special dry-powder extinguisher which has a yellow symbol on the label.

How will I choose the right one?

A <u>pressurized</u> <u>water</u> extinguisher is used to fight fires that occur in common combustible materials such as wood, cloth, and paper. Water extinguishers should never be used on electrical fires because of the danger of electrocution, or on grease fires or flammable liquid fires because of the danger of spreading the burning material around. Water extinguishers weigh about 20 pounds and should contain at least 2½ gallons of water. Any less than that is not going to do much firefighting.

<u>Dry-chemical</u> extinguishers work on Class B fires and Class C fires by smothering the flames. These extinguishers usually weigh about 10-15 pounds and should have at least a 2-A rating (on a scale of 1-10) The higher the rating, the greater the firefighting capacity.

<u>Multi-purpose</u> <u>dry-chemical</u> extinguishers work on all fires. These, too, smother the flames, but they are filled with a different type of powder than simple dry-chem extinguishers. This may be the best choice because you don't have to worry about matching the right extinguisher to the right fire (if there is such a thing as a "right" fire). The extinguishers are lightweight and easy to use.

All <u>inverted</u> <u>type</u> extinguishers, including the foam, soda-acid, and loaded-stream types are no longer made in the United States. They are ineffective and very dangerous if you don't use them right, or inspect and maintain them regularly. Don't buy one second-hand, and if you already have one, turn it in for scrap and buy another.

Where will I put it?

Install one in your kitchen at belt level so that it's easy to grab in a hurry. Mount it away from the stove so that you don't have to reach through the flames to get the extinguisher. If you're concerned about your decorating scheme, you may mount it in a closet or cupboard, as long as everyone (and guests or babysitters) knows where it is.

Also put one extinguisher in: the garage, attic, basement, and office or workroom. It's also a good idea to have one in your car or pick-up truck.

How much does one cost?

About $25, depending on the size and type. Buy the kind that can be refilled so that you pay only once for the cannister. Refilling is inexpensive.

How do I take care of it?

Check the pressure guage to see that it is within the safety range specified by the manufacturer. If you have a multi-purpose extinguisher or a dry-chem extinguisher, shake it to loosen any powder that might have caked together inside. Inspect the container to be sure no one has meddled with it and that there are no breaks in the container. Have it refilled or recharged whenever necessary. Call a professional extinguisher service company.

How will I use it? Does it "kick"?

The contents are under pressure, and there may be a slight jolt as you start to spray, but not enough to knock it out of your hands. Remove it from it's mounting bracket, pull the safety pin, point the nozzle at the base of the fire, and squeeze the trigger mechanism. Spray towards the base of the fire using short circular motions rather than a long frantic blast. Be conservative. The extinguisher holds only enough water or chemical to last a few seconds.

Why and where should I have one?

To protect yourself, your family, and your personal property if a small fire should occur. When they're used according to the directions, they can keep small fires from getting out of control, they can provide you with an escape route through a small fire, and they can let you fight a small fire while you're waiting for the fire department to arrive. **Always call the fire department whenever you have a fire.** Call them before you try to fight it yourself, because if you're unsuccessful with the extinguisher, or if the fire is too big for you to handle, you may lose valuable time. Fire moves fast. It can consume a whole house in minutes.

OTHER PROTECTION

A few of the practical things that are available include: poolside hose cabinets that draw water from the pool in case of a brush fire or roof fire; wall-mounted cabinets that hold a fire hose and an extinguisher; and low-cost home sprinkler systems. In fact, some cities and counties in the country now require sprinkler systems in all new-home construction. When remodeling, you should seriously consider a sprinkler system. They've been extremely effective in business and industry. Now that the price is right, it may be an excellent investment for your home.

If you feel that you need any of these items, a fire protection company or safety specialist from your fire department can help you.

9

General Safety - Part 2

Use your H-E-A-D
(Home Escape Alive Drills)

In the United States, home fires kill almost 6,000 people a year, and injure almost half a million more people. You can prevent problems from happening with your kerosene heater by following the guidelines in this book. If a fire or burn accident does occur in spite of all the precautions you've taken, you can still survive if you have early warning and an escape plan. An escape plan will work for you and your family if you know what it is and if you practice it.

USE YOUR H-E-A-D when fire strikes in your home. The rules for USE YOUR H-E-A-D are simple, and the plan will work if you rehearse your drills. Your safety depends on quick, clear thinking and action. Fire spreads rapidly. You may have only seconds to get out, so have a plan in mind.

Escape from a burning home in the daytime is not too difficult. But escape from a dark, smoky, burning home at night is an entirely different story. Here are the rules:

Draw a picture of your home.
Discuss it with the whole family.

When the family is together, draw a diagram of your house on a big piece of paper. Use heavy markers so that everyone can read it. Diagram each floor separately if your home is more than one story.

Two escapes from each room.

During a fire, the normal escape routes may be blocked. In your diagram, figure two ways out of each room. Draw big red arrows to the window or door exit for each room.

Walk through the plan.

If you've chosen a window as an escape, make sure that the window can be opened by the person who uses that room, and then make sure that the family member is capable of crawling through the window. If any member of the family needs help to escape, figure out which other family member is responsible for helping.

Two-story house, have ladders or ropes.

If your second-story bedroom window is not over a patio or porch roof, install a ladder or knotted rope. With a ladder, don't just buy it and put it on the top shelf of the hallway closet. Keep it on the floor right under the window, or maybe on the floor under the head of the bed. With a rope, get a piece of 3/4" nylon rope that's long enough to reach from the floor of the room, out the window and to the ground. Then tie a knot in it every 18 inches so that someone can shinny down it easily. Then bolt it to the floor in front of the window with an eye screw. Coil it up on the floor around the screw.

Who calls. Who counts.

Designate one member of the family as the person to call the fire department. Designate one more person as the "nose counter." If your family has only one adult, that adult takes both jobs. First get outside and count noses, then use a neighbor's phone to call the fire department.

Pick a meeting place.
Take a walk outside and choose one place where every member of the family must go to report for the "nose count." If everyone goes to the same spot, you'll know right away who is safe and who still needs help. Many "trapped" victims turn out to be not trapped at all, just misplaced or disoriented after getting out.

What to do when the alarm goes off.
Close your bedroom door if you sleep with it open. Feel your door if you sleep with it shut. If smoke, flame, or heat are present, keep the door closed. Don't open it to peek. Stuff a blanket in the bottom crack of the door. Keep low, crawl along the floor to the other exit, get outside and meet.

Once out — stay out!
No animal, treasure, jewelry, or picture album is worth the loss of life. Once all members of the family are at the "nose count" headquarters, stay there. If everyone does not show up for "nose count" be prepared to tell firefighters where the trapped person is. Don't go back in yourself. Firefighters have breathing equipment. You'll die from the smoke. The person inside may still be all right if he or she is following the shut-the-door-and-stay-low-and-crawl rules. Instruct kids to "hug the tree" or "make like a rug" on the ground so that they understand just how important it is that they stay right where they are without moving an inch.

What to do if trapped.
Go to the window and wait to be rescued. Don't open the window if smoke or flames are licking

up at you. Rescue procedures are very thorough and firefighters will find you. Tell children never to hide in a closet or under the bed. The fire can burn right through closet doors, so tell them that they are safer if they wait by the window for help if they can't get out by themselves.

NOTE: It's a good idea to equip each adult and older child in your family with individual flashlights (and fresh batteries) to see in the dark and to signal if trapped.

ANOTHER NOTE: Bars on the windows and key-operated deadbolt locks can trap you inside and can prevent rescuers from reaching you. Whenever you're planning a home security system, keep in mind that you don't just have to keep the burglars out. You also have to be safe. Use thumb-latch deadbolts that you can twist in the dark and smoke (even though you shouldn't be using the door to escape if there is a safer way out your bedroom window).

Almost once a month you can pick up a major metropolitan newspaper and read an account of someone who was trapped behind bars in their own home and perished because no one could get in to save them. If you must buy bars to protect your property, make sure you get the ones that can be released from the inside, and make sure that the opening is large enough for you to crawl to safety.

ONE LAST NOTE: When you involve children in a home escape plan, do it with love and concern, and make them feel like they're part of the decisions in the plan. Let each child draw his or her own escape-plan picture and then encourage them to hang that picture proudly in their room.

10

General Safety - Part 3
How to summon help
First aid for burns

First a three-page refresher course on how to call for emergency help. Panic does strange things to people. Emergency services across the country have reported all kinds of unusual calls for help. An incomplete call means you lose precious time. Inaccurate information means the same thing and could cost someone's life. Sometimes people get so caught up in the frenzy of the emergency, that no one calls for help until it's too late.

Always call the fire department. Dispatchers would rather receive half a dozen calls about the same fire than no calls at all.

When reporting a fire emergency, follow these steps:

1. Call the fire department directly. Know and memorize the number. Write the number on a sticker, and put a sticker on every phone in your house. Also write your address on a sticker and put it on every phone as well. Why? You might forget where you live, any guests may not know

your address well enough to recite it in an emergency, and babysitters may not know your address.

2.   If your city or town uses 911, use the number and be prepared to state the exact nature of the emergency when a dispatcher comes on the line.

3.   If you're in a strange place or stopped at a pay phone and you don't know the fire department number, don't stop to look it up in the phone book.

Call the operator and tell her or him the following:
    A.   Your location.
    B.   What city you're in.
    C.   The phone number of the pay phone.
    D.   The address on the phone if listed.

4.   When the fire dispatcher answers:
    A.   Give the address of the fire.
    B.   Indicate an apartment number, the correct space in a mobile home park, what floor of a multi-story building.
    C.   Indicate the closest cross streets.
    D.   Tell what your phone number is.

5.   Give a short description of the problem, what is burning.
    A.   Car on fire.
    B.   House on fire.
    C.   Barn fire.
    D.   Truck fire.

6.   Also give any more important information if you have time, such as:
    A.   Car on fire, it's in the garage.
    B.   House on fire, heavy smoke, folks trapped inside.
    C.   Apartment fire in downstairs, folks asleep upstairs.
    D.   Barn fire, livestock has been evacuated.
    E.   Truck fire, loaded with gasoline.

7. Stay on the line if you can in case the fire dispatcher needs any more information.

Many fire departments in the country also respond to medical emergencies. The procedure for making the call is the same, although you may need to give a little more information about the exact nature of the emergency.

When you call in a medical emergency, it's vital that you stay on the line. The dispatcher may be able to give you one-step-at-a-time instructions in emergency life-saving techniques.

When firefighters and medical-aid personnel arrive at the scene, stay out of their way, but stay close enough so that if they need your help you can give it.

FIRST AID FOR BURNS

Now, because accidents can happen to even the most safety conscious, just a few more pages on first aid for burns. You need to know only three things about emergency first aid for burns.

COOL WATER

CLEAN SHEETS OR TOWELS

COMPETENT MEDICAL CARE

COOL WATER

Skin continues to "cook" after it is burned. Cool the burn immediately with standing tap water or a very gentle dribble from a hose. Do not pack in ice, rub with ice, or stick under a wildly running faucet. Cool the burn as soon as it happens.

P.S. This also works to ease the progression of sunburn. By sitting in a tub of cool water, you can ease the pain of a sunburn.

CLEAN SHEETS OR TOWELS

Keep a burn clean until you can be seen by a doctor or until emergency help arrives. The only thing you should put on a burn besides cool water (or in place of cool water if none is available) is a clean sheet or towel.

NEVER spread butter, grease, oil, vaseline, or any other ointment on a burn. Never wrap a burn in gauze. If anyone is going to put any ointment or any dressing on a burn, it should be a specially-trained burn doctor, nurse, or technician. Butter or grease only helps to hold in the heat and "cook" the skin some more. It makes the burn worse, not better.

COMPETENT MEDICAL CARE
IMMEDIATELY!

Burn care is a highly specialized field of medicine. The first few hours after a burn are just as critical as the first few minutes after a heart attack. A comprehensive burn center, or a hospital with designated burn-care beds, is where you should be treated if you've been severely burned, have suffered smoke or gas or heat inhalation burns, or someone you care about has been burned badly.

If you live in a metropolitan area, there is probably a burn center within driving distance. Call the county medical association for the location of the nearest comprehensive, accredited burn center. If they don't know, call the fire department, call the hospital administrator, or write to the American Burn Association. When you get the information, keep it in the front of your phone book and in the back of your mind so you'll have it when you need it. You need this information before you're burned, not 6 hours later. Care in the first few hours after a burn can mean the difference between life and death, or between crippling and amputation or functional use of all your limbs.

11
Conclusions - Recommendations

So, 80-some pages later, what's your opinion on kerosene heaters? Are they safe? What's your attitude on family fire safety and burn prevention? Is it important to you?

On page 4, I told you that I would draw some conclusions and that I would try to present all the information that I've been able to gather in as factual and fair a way as possible. My original statement still stands: the heaters as designed are safe. **But,** the safe operation of the heater depends entirely upon a safety-conscious consumer.

The three biggest issues in the controversy were: fires (and fire-related deaths); contact burns, and emissions that reportedly lowered indoor air quality to dangerous levels. The many studies that have been done indicate that the heaters by themselves do not present unreasonable fire, burn, or emission threats. Over and over the experts have stated to me, to government officials, to reporters, that the key to acceptance and safe use of the devices is consumer cooperation.

I'd like to digress just a paragraph or two to draw a parallel between kerosene heaters and automobiles. Both products have the capability of being very dangerous or very convenient, depending on how they are used.

When you drive a car, if you're safety conscious, you have first made a subconscious mental note of the risks involved. Before you start the car, you have a general idea of where you're going, you have put the right fuel into it, you know how it operates, and you know how to turn it off if you have a problem. You also wear your seat belt to protect against any sudden changes.

Correspondingly with a kerosene heater, before you light it, you know what you want it to do, you've put the right fuel into it , you know how it operates, and you have back-up protection and plans if an emergency occurs.

Once you're in the car, seat belt still fastened, you observe traffic laws, drive courteously yet defensively, and drive only when alert (not when sleepy or under the influence of alcohol or drugs).

Correspondingly, with your kerosene heater, you observe the safety precautions regarding refilling the heater, storing the fuel, and distance between the heater and things that will burn; you pay attention to the way the heater is operating and are alert to any sign of malfunction, bad fuel, or damaged wick; you keep small children and animals away from the heater; and you always plan for sufficient ventilation for the heater.

I saw a sign in the post office once: "Accidents don't happen, people cause accidents." Heaters and automobiles don't have accidents by themselves; **people** cause accidents with automobiles and heaters.

Let's briefly review the story and the conclusions:

FIRE HAZARDS
The single greatest fire hazard involves the fuel. A CPSC study calculated that for a family of four the risk of fire (per 9 million users) is one in 1,297 years. Only you can determine if you'll be the "one" or the other 1,296. There were over a hundred fires in the 1982-1983 heating season, so it is a real risk.

To minimize the risk from any fire/fuel accident, always use 1-K kerosene fuel -- **nothing else,** don't contaminate the fuel with any other fuel or with dirt or debris, don't fill or re-fill the heater in the house, don't store the kerosene in the house, don't store the kerosene where children or animals can spill it (or be poisoned by drinking it), and always wipe up any spilled kerosene.

The fire-safety record for the heaters is not that bad, but you are the controlling factor.

BURN HAZARDS
Most of the contact burns that have been reported happened to young children who fell against the heater. Either the youngster was running and fell, or was an unstable toddler just barely able to walk who stumbled into the heater. The temperature on some heater cabinets has been tested to as high as 700°F. When you consider that less than 5 seconds of exposure to hot water at 140°F. is enough to cause a 3rd degree burn, you can imagine the damage that would result from touching or falling into a heater that is four or five times hotter.

The second most common contact burn incident involves adults who stand too close to the heater and are burned from the radiating heat. This kind of injury is particularly common in older people whose skin is less sensitive to hot and cold. NEVER stand any closer to a heater than 3 feet, the same distance as anything else that will burn.

When buying a heater, look for one that has protective guards and grills. If you already own a heater that does not have these grills -- and if there are children or elderly people in your household -- buy or make a fence-like "cage" for your heater to prevent anyone from coming in direct contact with the cabinet. You'll still have plenty of heat, and no one will get burned.

EMISSION HAZARDS
Indoor Air Quality

As I said before, any heating or cooking device that uses a flame will use oxygen from the room in the burning process, and will release products of combustion -- contaminants, pollutants -- into the air. **Any heater, any cooking device.** In addition, any heating system or device that burns petroleum products will give off carbon monoxide, carbon dioxide, nitrogen dioxide, and sulfur dioxide, the same gases that pollute outdoor air all over the country -- some areas more than others.

To provide a little more perspective, a single cigarette produces 30 times as much of these pollutants by concentration.

The difference between a central heating system and a kerosene heater is that the central heating system vents these gases to the outside through a chimney. A kerosene heater "vents" these gases directly into the room it is heating. To minimize the indoor risks from these gases, you must provide for circulation of the air -- to other rooms or in from outside. The CPSC advises that you provide 10 square inches of opening for every 10,000 BTUs of heater size, the equivalent of a quarter-inch opening in a window that is thirty inches wide.

As of this writing, no one agrees on standards for indoor air quality. Some experts apply outdoor standards to indoor air, others use standards from the Navy's submarine service, some use standards set by OSHA. Complicating matters is the fact that many new or remodeled homes are built to "energy efficient" specifications which also lead to poor indoor air quality. A "tight" house lets out no expensive heat, and also keeps in all contaminated air.

Heating systems or devices aren't the only sources of unhealthful indoor air. All matter -- all things natural or man made -- gives off gases as it warms. Even the normal day to day temperature of a house causes paint, carpet, furniture, and all things in the house to vaporize to some extent. Before energy conservation, when homes leaked a little bit, the indoor air escaped and the outdoor air invaded. This air exchange kept indoor air moving and at an acceptable health level (unless, of course, the outdoor air was even worse).

ARE KEROSENE HEATERS SAFE?

I think so, the long and expensive studies appear to prove so, 9 million+ consumers believe so or they wouldn't have them in their homes. People don't intentionally buy things that they think will kill or injure them, and then intentionally use them in a way that's sure to kill or injure themselves.

ARE 9 MILLION+ CONSUMERS SAFE?

I hope so. So does the CPSC, so do the heater manufacturers and distributors, so do members of the fire service and the burn care profession. Your intentions -- your full attention -- with regard to safety will make the difference.

To borrow a line from Smokey the Bear: "Only you can prevent" any problems with your kerosene heater.

Resources

To find out the status of any developing research, or to report any difficulties on any product:

CONSUMER PRODUCT SAFETY COMMISSION
Office of Program Management
5401 Westbard Avenue, EX-P
Washington   DC 20207

*   *   *

To find out the location of the closest hospital offering accredited acute care for burns:

AMERICAN BURN ASSOCIATION
Duke University Medical Center
ATTN: Joseph A. Moylan, M.D.
Box 30
Durham   NC 27710

*   *   *

At this writing, some heater manufacturers are moving their corporate headquarters. Please refer all questions to the heater industry's organization for the current address of the company that manufactured the heater you own. The reference librarian at your public library can assist you, also. He or she has available several books which list manufacturers and products.

NATIONAL KEROSENE HEATER ASSOCIATION
First American Center #15
Nashville   TN 37238

Acknowledgements

Very special thanks to Jim and Joan Groody of Sun • Wind • Fire, an independent heater sales and service store in Lambertville, Pennsylvania. Each consumer deserves someone like the Groodys to guide them through the choosing and using of their kerosene heater. Their technical knowledge is exceeded only by their willingness and ability to share that knowledge unselfishly.

Thanks also to the kerosene heater team at the CPSC, to members of the fire service who always asked provocative questions to make me think hard about the issues and dig harder for the answers, to the tireless 3-person staff of the NKHA, and to all agencies and firms who assisted with technical information or photos or drawings.

By no means last or least, thanks to the staff of Huntington Beach Design Group who never fail to come through with typesetting, graphics, design, or camera work to make my work look better. And they always do it cheerfully, just what I need when deadlines are closing in much too quickly.

And Gary, as always, a remarkably patient man.

About the Author

Peggy Glenn is an award-winning professional writer specializing in home and family safety and topics of interest to people who run home-based businesses. **Kerosene Heaters: A Consumer's Review** is her fourth book. She has also acted as editor or consultant on five additional books. In addition, Glenn has written for many magazines and newspapers on home safety or home business (or home safety as a business, or home business safety).

She first became interested in family fire safety in 1975 when she married a firefighter -- a fire prevention and investigation specialist. Since 1981, she has been active in local and national organizations whose aim is to promote fire and burn safety through prevention and education.

Peggy Glenn lives in Huntington Beach, California, with her husband, Gary, and their family.

Order Form

TO: Aames-Allen Publishing Co.
        924 Main Street
        Huntington Beach CA 92648

FROM: _____

_____

_____

Phone: _____/_____

Please send the following to me immediately:

___ **Kerosene Heaters** @ $3.75   ___.___

___ **Don't Get Burned!** @ $7.95   ___.___

___ **How to Start and Run a Successful Home Typing Business** @ $14.95   ___.___

___ **Word Processing Profits at Home** @ $14.95   ___.___

Shipping: $1 first book,
           50¢ each additional   ___.___

Californians add 6% tax please   ___.___

TOTAL   _____.___

___ My Check is Enclosed
___ Charge my ___ Visa® or ___ MasterCard®

Card Number _____

Expiration Date _____

Signature _____ * * * * * *

COOL THING

THE BEST NEW GAY FICTION
FROM YOUNG AMERICAN WRITERS

COOL THING

THE BEST NEW GAY FICTION
FROM YOUNG AMERICAN WRITERS

Edited by
Blair Mastbaum and Will Fabro

RUNNING PRESS
Philadelphia · London

Cover Designed by Scott Idleman

Interior Designed by Jan Greenberg

Running Press Book Publishers

2300 Chestnut Street

Philadelphia, PA 19103-4371

ISBN 978-1-60751-361-2

CONTENTS

EDITOR'S STATEMENT
(Will Fabro)

GENTRIFYING THIS GHETTO

It has been remarked, jokingly and with reason, that both the artistic and gay communities are catalysts for gentrification. Displaced from more "desirable" areas due to the demands of real estate, economy, or cultural hegemony, both marginalized camps flee to less trodden, "off the map" areas which then become bustling sites of diversity, of creative expression. A sense of renewed opportunity. After a few years, this opens the door for the dominant culture to swoop in, displacing them once again to further-flung territories along with the neighborhood's original denizens. Vicious cycle.

I can look out my window and see this at play: stroller-pushing Puerto Rican families walking past the new gourmet coffee shop, chainsmoking white hipsters on the patio with their Macs and silly hair. Homeboys order their chicken wings from hole-in-the-wall Chinese joints while down the block there are spindly young men who look like they're competing for a fashion-related reality show, heading from the organic market to their lofts two blocks away, groceries in one hand and art supplies in the other. On the street you can hear salsa music blaring out of the windows of one apartment, and across the way thrash metal competes to be heard. A car with its windows down speeds by, blasting reggaeton that still sounds clear three blocks away. When I walk into my corner deli, the Saudi gentleman automatically reaches for my brand of cigarettes and asks me about the rise of Islam in the Philippines and have I ever been? Yes, I say, I still have family there.

Completely ignoring the troubling race and class issues that spring forth from gentrification, the neighborhood feels right. Like home.

Writing gay fiction is a double-edged sword—on one hand you can almost guarantee a built-in audience of devoted and passionate readers; on the other, you run the risk of being assigned to a niche, of permanently residing in the "gay ghetto." You can write sexually frank provocations, or create more traditional pieces of literary craft, or concoct genteel Boyfriend Stories, but you will be stamped as a "gay writer" rather than simply a "writer," thereby necessitating a marginalization from the mainstream literary community.

Certainly there are gay writers like Edmund White and Michael Cunningham who have rightfully staked a place in the upper echelon of American literature. There are bonafide cultural phenomena like Davids Rakoff and Sedaris. And then you have writers like Dennis Cooper and Kevin Killian, whose uncompromising work makes them less palatable to the culture at-large, yet they are nevertheless revered as paragons of iconoclasm and daring.

The writers featured in this anthology benefit greatly from the influence and trailblazing of the above authors, and many more. The editor Don Weise (of the now-defunct Carroll & Graf) approached Blair Mastbaum and me with this project, under the idea that it would portray a new sense of the gay male voice—specifically the young. Mr. Weise should be commended for his dedication to the cause; after a feverish period in the 1980s in which gay and lesbian writers cultivated both a sense of community and an audience, much of these inroads would come to be decimated by AIDS. Since then, Mr. Weise has been at the forefront of reawakening that sense of community, of fostering the notion that gay fiction is not only alive and well, but exciting and new. Blair and I thank him, and also Greg Jones of Running Press for continuing with the project.

What inspires about the writing contained in this anthology is in how the aspect of sexuality is attacked, deconstructed, manipulated, or even dismissed. Gay sexuality here is not an internal struggle; the coming-out narrative, a necessary and important archetype of gay fiction for the past thirty-something years, is almost a relic to these writers. It is instead a matter of fact: wildly external and exuberant, sometimes a banal secondary issue, at other times barely acknowledged or even existent, and in a few cases so confrontational as to rub your face in its flesh, hair, and musk.

If the original intent was to have an anthology that would showcase a predominant voice of young gay men,

the stories contained here bristle at any such reduc-
tivism. Because there is in fact no singular voice, no
homogenous mindset. There are myriad voices portray-
ing a tale of fractured community: herein are stories of
domestic longing, sexual violence, provocative confes-
sion, spite and redemption. The protagonists are alter-
nately beautiful and ugly, intelligent and inarticulate,
contemptible and lovable—sometimes within the same
story. In short, this collection of writers are relaying tales
of humanity in all their various colors and contradictions. And for all their dissimilarity, they are brought
together under this umbrella of "gay fiction." If they
must be ghettoized, they are at least working with maximum vibrancy and passion in an area that allows it, displaced as they are.

In your hands is a goddamn rainbow, and the view
from my window looks pretty good.

— Will Fabro
Brooklyn, NY

EDITOR'S STATEMENT
(Blair Mastbaum)

Sadness, a snowstorm, a gleaming gun, quiet New York streets, a pickup truck, a beach holiday with Brad Pitt, a dead ex-boyfriend, a teacher's inner voice, a porn shoot, a climbing monkey boy, a sonic adventure, a buck in Manhattan, an eco-terrorist, a dead mother on a psychic hotline, thug rape, domestic unhappiness, a Pabst tall boy, a pervert, New Year's Eve, the Puente Hills Mall, real estate, a ghost, a laundry room, an abandoned KOA campground in the swamp.

The only thing that seems to make today's homo writing similar is its dissimilarity. I have nothing in common with most of these people, or these experiences. And these stories surely don't portray some happy group. To me, they're all about searching, and not always finding, some basic things like love and sex and food and money and a sense of belonging. But they're also about running away from things, hiding, dodging intimacy and domesticity.

I hope you like these works of fiction. They most certainly illuminate what it's like today to be a young and gay and passionate about expressing experiences in the written form, and I think that's pretty rad.

— *Blair Mastbaum*
Portland, Oregon

MR. MIGS MAKES ME CRAZY

by Miek Coccia

Mr. Migs drinks 40s and likes to climb things. Any things. I have a picture of him sitting wedged at the top of the hallway, his head pressed up against the ceiling, smiling down at me. He was bored he said, Just Bored, and so he went up there to—he doesn't finish the sentence, just smiles instead. Crazy he is, I call him Crazy Migs but not to his face. He makes me crazy too. I snap a picture with my cell phone and tell him to come back down, come to bed, it's time.

He makes me crazy when he's near, but mostly when he's away; away so much it makes me crazy. Gets me crazy looking for him online when he's gone—Gmail, MySpace, Yahoo, and AIM, looking for a signal he's out there able to see me, signed in, all by himself and not keggin' it up with friends, climbing over fences for the fuck of it, hiding in some tree. I wish I could call but he doesn't have a cell phone, doesn't let me call his house cuz his mom doesn't even know he likes boys.

Oh Migs u make me crazy u do.

I met Mr. Migs on Craigslist. He had just quit his job, his UPS job, and was looking to make some quick easy cash. He wouldn't have sex for money though, he was very strict about that but I thought he was just trying to avoid the law, in case those Internet cops you always hear

about were around, how embarrassing to tell his mother. He'd only have sex if he was attracted to you he said,
and he wasn't attracted to 40 yr old men he said, but if
you wanted to pay him to have dinner or sit around or
something that'd be cool with him, that's what he said,
it'd be cool. I'm only 26 so I sent him an e-mail, two e-
mails, tried to figure out what the ad was really trying to
say. His pics were way cute and I'm sooo into 19yo's, I'd
let him name the price, and just hope it was under 200,
200 is my maximum, though I could maybe spend a little more if he was willing to sleep over, let me rub his
tummy to sleep sleep. He was that cute, that 19.

***

Mr. Migs hasn't messaged me for two days now; it's been
four since I've seen his sweet face, at the diner where we
meet cuz I don't have a buzzer, his big bright smile saying "I can't believe you waited for me" over and over,
after he fell asleep on the subway, four o'clock in the
morning and who remembers how many bottles of wine,
the cheapest thing he could buy with his unemployed
wallet, all his friends and their unemployed wallets,
stealing money from their mothers. I love it. He woke up
in Coney Island.

   "I can't believe you waited for me," he said, he was
thinking all the way from Coney how to break into my
apt, he wasn't gonna take the train so far back home so
he decided he'd kick the front door to my building in, go
out the back and knock on my window, crawl in my win-

dow if I didn't wake up, break through my window if it was locked, steal some money from his mother to replace it. But I waited for him. "I can't believe you waited for me." And then he says it again, and I kiss him while we're walking under the BQE. It was kinda romantic, I thought he might think it was romantic. When we got back I showed him how to break in without breaking anything . . . the basement door is always unlocked.

But that was four days ago, him so proud of me, my waiting for him, and his holding my hand on the subway to work in the morning.

Four days and he's in and out and two days with no message now, but I see his MySpace, his AIM, and his Gmail online. I know every minute he's been online and off, I've read his Away messages and know where he's been, though "MMMM Donuts . . ." took six hours, how many donuts can one person eat? And he's online now but it's been 12 minutes and he hasn't msg'd me.

I get a bit pissed, but I won't send him another message, I won't chase after him, pathetic me won't go crazy push him away, I always scare them away I know but not this Mr. Migs cuz he's one to hold on to I know that, so this time I know I'll do better, won't scare him away no.

But it's just so hard to pretend not to like him so much as I do, to not send him an e-mail an IM a profile comment at every hour gone by that I miss him, wonder when I'll get to see him again, hold his head in my lap while we watch the next stupid movie On Demand.

And I know this, I know, but it's just so hard. It's not

like we have anything to talk about anyway. But I love so much to listen to him, to listen to 40s and climbing things and all those things he talks about so well, fascinates me with, I want so much I wish so much I got to videotape him.

***

Mr. Migs makes me crazy cuz I shouldn't like him to begin with. We have less than zero things in common, I know this, we have nothing to do with each other's lives, the way we are and the way we do things. But we want to, we want so hard to be each other. Does he want to be me? I want to be him I think. I love the way he has nothing to do with gay culture, hangs out with breeders all the time and gets in fights, starts fights cuz he doesn't like someone's shoes or something. I wish I was tough like that, young like that. I wish I knew how to climb things. And he tells stories so well without sounding stupid. I always sound so stupid, I always wonder if he thinks I'm stupid. Christ. Christ, he's only 19.

The first night we met we met at the diner, estimating the time it would take him from Fresh Meadows, the only thing open near the subway; I'd buy him dinner I said, like the ad said. I got to the diner at 2:20 like he said, walked in but didn't see him, that boy in the pictures, and I got scared for a moment I was throwing my money away, to some cute boy who wasn't so cute anymore, so not cute that I couldn't even recognize him, cuz that happens sometimes with these Craigslist boys. I went to the counter to buy some tobacco and when I stepped around Mr. Migs was there, he wasn't smiling or anything but I knew it was him cuz he was 19 and so cute and looked just like his pictures, a little shorter than I thought but I like that, I have little ceilings.

He wore green and so I was like, "Hey, so you hungry? Wanna get some food?" and he was like, "Nah, not hungry" and so I was like, "K, so you wanna just go back to my place, get some beers or something?" and he was like, "Yeah, that's cool."

And I didn't have any beers at my place so we had to go across the street to the grocery. I was a bit nervous; he didn't seem charmed or anything, interested in me at all, maybe like he just wanted to get wasted, let me suck him off and then take the cash and go home. In the grocery store I ran into this fat girl I met at a party a few weeks before. She's really really fat. When I came out I was like, "Hey, did you see that girl I was talking to, the really big girl?" and he was like, "Nah, I usually don't pay attention to fat people." And I was just like, omigosh, this kid is rad. I mean it's not like it's so cool to not pay attention to fat people, but just that he'd say it, knowing me for five minutes, that's fucking rad. I wanted him so much there; I knew he was nothing like me.

We went back to my place and I didn't have a couch yet, just a big chair and a little chair. I offered him the big chair but he insisted I take it, that he didn't care. So I was like, "Cool."

I can't remember so much about what we talked about, just that we smoked weed and I don't normally smoke weed, so I wasn't feeling very well and not even so sexual, so I went to the bathroom a few times and took little nibbles on the Viagra I keep in my keychain, just in case. I was a bit frightened too, I get that way with weed, and I was happy that I remembered to hide the electronics and shit, cuz you never know with these Craigslist boys and god he doesn't even seem gay at all I remember thinking, like what if he was one of those gaybashers or something and was just gonna like, kick the shit out of me like they did in the old days, call me fag and take all my money, but I hid my stash really well too, and even made a little stash just in case he wanted to beat it out of me you know, cuz you never know with these Craigslist boys.

But yeah, he was cool, he showed me which cartoons were cool these days, how to use my remote to watch things On Demand and shit, I had never bothered to look that stuff up. I even thought you had to pay extra for it. I remember thinking though that still he wasn't into me, that maybe something would happen tonight but probly not much, I'll suck him off a bit, he'll get bored, and then I'll feel really bad about it, about him being so bored, so I'll just jerk myself off and Get It Over With.

I felt a bit sick because I'm bad at smoking weed and I smoked too much cuz he kept pressuring me. But he felt bad I think, about pressuring me and then me getting a bit sick, and so he was like "Let's go lie in bed, watch a video or something," and he picked out *Donnie Darko*. We got the little TV to work and got into the bed and I remember it was really sweet, how he pulled me up to cuddle into him, pulled my head into his chest and my arm on his stomach. We laid around like that for a bit, me stroking up and down his torso, his little bits of hair down there at the bottom of his tank top.

Time went by and I got a bit braver, wandered down his boxers onto his leg, trying to accidentally rub against to see if he was hard or not, if he wanted me to go down there, and he was, he did, so I went there and things happened, they happened really nice and we kissed a lot and we sucked a lot and we rubbed all over and it was good and I knew he wouldn't let me fuck him, I just felt it, so I asked for his cum all over my face, to feel it there and it felt so good, so good I tasted a little bit before putting him back in my mouth yeah, which is bad and unsafe but it felt right, and I had the weed so I let myself. And then the movie was ending and that was OK and so we hooked up into each other to sleep.

In the morning we woke up and we did some stuff again, but he was pressing back against me and I knew he wanted it inside him, in the morning, so I gave it to him, and fell asleep a bit more that way before I had to wake up for work. I got my money's worth I thought, I never really feel that way with these Craigslist boys, but I did this time, no matter what he wanted I got my money's worth, I learned so much, a new experience. I still didn't know how to bring that up to him, the money, he never said and I asked I did, a couple times online the night before, and he avoided it and so I was like well let's just see what happens and you can let me know and he was like, "Whatever."

I took a picture of him in my bed, his boxer shorts in the light were covered with Teenage Mutant Ninja Turtles and that melted me a bit, I had to take a picture, but I really melted when I went into the shower, clean myself off and get ready for work, and I turned around and he was sitting there, just sitting on the toilet watching me, my scrubby dub dub, still no smile but curious, watching me, making me crazy already this Mr. Migs.

After getting dressed I thought OK I have to do the money thing now and so I asked him, "So hey, uh, like, so I want to help you out with some money, what were you thinking," and he was like, "Nah, you don't have to pay me, I said I was into you, it's cool." And then I was like, "No, it's no big deal, really, I mean you're unemployed," and he was like, "It's cool, I still have some money," and I was like, "Well can I at least just give you a hundred bucks or whatever, really, it's no big deal," and he was like, "Nah," and so I dropped it.

We were walking to the subway and I tried to bring it up again, but he still wouldn't take it and then he took my hand and held it down the street and so I was like, omigosh, and I dropped it again, the money thing. Oh Migs, you got me soooo crazy. I hate this.

Mr. Migs drinks 40s and likes to climb things. Any things. I have a picture of him sitting wedged at the top of the hallway, his head pressed up against the ceiling, smiling down at me.

He makes me crazy when he's near but mostly when he's away; away so much it makes me crazy. Gets me crazy looking for him online when he's gone—Gmail, MySpace, Yahoo, and AIM, looking for a signal he's out there able to see me, signed in, all by himself and not keggin, it up with friends, climbing over fences for the fuck of it, hiding in some tree. I wish I could call but he doesn't have a cell phone, doesn't let me call his house cuz his mom doesn't even know he likes boys. I shouldn't see him again anyway, not yet, a few more days maybe. Grah.

Oh Migs u make me crazy u do.

*****

I remember walking the streets til midnight on a New Year's Eve. Another year gone by and I still didn't find anyone, anyone to do the things I wanted to do with me. I remember snow on the ground and snow in the air and everyone up inside, in big groups on all the different floors of the city. I remember the sidewalks so empty and so often I had the fresh snow to myself, to make my prints in and in any way I wished to make them, my prints—to drag my foot a bit sideways, to waddle like a duck, hoping my True Love would come along, recognize my steps, his steps, and follow up behind me, find me. My True Love, Quack Quack, is out in these streets I thought.

And I remember when the New Year turned over, and all these big groups on all these floors made big loud crashing noises out into the streets, my streets, with their woos and waas and little tiny pieces of plastic spinning around, back and forth, beating all of them out.

And I wanted to cry and I tried for a minute, to cry for another year gone by, a year without peanut butter cookies, going to see what Central Park was all about, that space on the river looking into Manhattan, all those things I saved for Two. I tried but it didn't seem to work very well, this crying thing, right then, and so I just pointed my footsteps back home, Quack Quack, get comfortable, find some sleep tonight and forget about all of this, tomorrow, this next New Year to me, Change. Sigh.

***

I'm bored.

There are monkeys on TV and they remind me of Migs, because they climb things. But the TV is muted and so I don't really pay attention to them, they just flutter around at the corner of my eyes every now and then, look down at me Just Because. iTunes is shuffling into another sad song and I'm clicking away at YouTube, following a cat climbing in and out of a bottle to some kids throwing bottles at a dead car to some kid throwing tacos at McDonald's employees. I fucking love it, this Internet thing, I can't wait for the rest of the world to just crumble away from neglect, leave us to our boxes and screens all wired up for whatever we want whenever we want it. I want Migs now but can't yet and I hate that.

It's been four days since we've chatted, over a week since I've seen his sweet face, saying goodbye outside the diner where we meet, because I don't have a buzzer and he doesn't have a cell phone and it's romantic that way I think, it's like when people were bohemian way back when. Sigh. I'm bored again. I click over to AIM to see if he's in yet, then over to MySpace, see if he has any new comments, to Gmail, Yahoo, then back to AIM because you never know, you look away at the monkeys on TV for just a second and you never know, he may have signed on. It's happened before, I've missed him and gotten wrapped up in whatever else there is and then you remember, finally, to check and he's been there over thirty minutes. I hate that. I can't wait for the New Computers, everything on one screen, no windows to click through, everything alerting you when anything happens, not just Forward and Back but Something Else, something more Real. Christ, it gets me so pumped, I fuckin' love it.

I type "monkey" into YouTube and watch a clip of a monkey playing with a puppy's penis. It's stupid but kinda funny.

My phone buzzes a new text message in. It can't be Migs though cuz he doesn't have a cell phone. I know it's just Alex saying he's bored again, he checks in every few hours to let me know. I get up cuz I left the phone on the couch, and it is Alex, and he's bored and I laugh out loud for a sec like I always do cuz his timing is always really perfect. "Cool," I reply, then wait a minute. He doesn't write back so I add, "Yeah, me 2." A few seconds pass and he hits me back, "Cool." LOL. Me and Alex think we're on to something, that Bored could quite possibly be the new Cool. We almost swear by it. I don't know I mean, it could happen I guess.

***

I remember my way home that night, unable to party with the hip kids and so sad on myself I couldn't cry, walking through snow the quickest way home, enough of this. Looking down and avoiding the people going by, I see a little spot up ahead, and find a baby mouse frozen in the snow. Little mousey-poo, so cute and young and frozen in mid-stride, trying to cross the sidewalk, I'm not sure why or where but he's so cute and so frozen in mid-stride. I remembered those theories that you could freeze people after they die and one day in the future the technology will be so advanced that we might be able to bring them back to life. I don't know why but I decided to keep the mouse as a little friend.

***

Another hour goes by and no sign of Migs and I start to get a little crazy inside cuz I just wanna chat and ask how his day was and tell him about my week so far and that I picked up a book for him even though he doesn't read but I think he'll like this book it means a lot to me.

Alex sends another text message but I left my phone on the couch again and just don't feel like getting up right now—I'm clicking and it feels good and so I decide to wait till the end of this video clip but three minutes is too long, I hate when these people post videos that are longer than a minute, unless they have music in them, and so I say fuck it and get up off my ass. But hey, cool, it's not Alex, it's Robert, I haven't heard from him in over a week and yeah yeah, I reply, I'll be right there.

I change my away message to let Migs know I'll be back soon, wash my face and fingers and head out the door.

KYLER AND WOLF-BOY

from *The End of New York City*

by Benji Morris

KYLER

He says his name is Kyler, but no one believes him. It's been rumored that Kyler is his porn name. He's ecstatic, walking down Fifth Avenue. His hair hides his eyes and he's wearing a black hooded sweatshirt, dark blue jeans, and silver spray-painted Converse sneakers. He's got a blue bandana in his back pocket. He's still wearing a fucking wallet chain even though most kids stopped wearing them ten years ago.

He always does random wild acts, like kicking parked cars and breaking out their headlights, like drinking crazy amounts in very short periods of time, like stealing designer clothes from shops, like jerking off for some old pervert for some amateur porn website.

He always says, You have to do something wild if you're from Iowa.

He moved to New York City a month ago, and he still doesn't have an apartment or a room. He's staying with the 52-year-old who films him with morning wood, films him taking showers, films him drinking beer and smoking cigarettes. He even shaved his pubic hair for the old dude, even though it itches horribly and looks lame in his opinion, and he would know.

Back in Iowa, he was the catch of the town, the county, maybe the tri-county area. He was the prom king, the homecoming king. Then he was dethroned when his girlfriend caught him taping himself jacking off for the guy that he's now living with.

She would have been fine with it, she said, if it was for a regular, straight erotic web site or magazine, but since it was for other men, she could never look at him again.

He used her rejection to do something he'd always wanted to do. He bought an old white van and drove to New York City. He always had crazy eyes. He always longed for freedom like old Western pioneers. He was too wild for his town of tractor auctions and enormous Wal-Mart parking lots. He never could have run the whole maze—all the way from prom king to father to grandfather to retiree in Western Florida. He'd have set the house on fire. He would have resorted to bizarre secret sex in truck stops. Or something. Or worn wigs. Well, not the wig.

He can't envision himself wearing wigs. His real hair takes too much work. It's got the perfect just-woke-up fucked-up-ness about it right now. He could be a real model, but he doesn't know who to talk to about it, or he doesn't try to find out.

The thing about Kyler is that he feels he has nothing to lose. The truth is that he doesn't.

He has no family. He has no friends. He's basically being kept by a pervert pornographer who acts like he cares about him but always invades his privacy, wakes him up at odd hours to jerk off or let him lick his ass-hole. It's tedious and demeaning, but he can't do any-thing about it right now.

He steals money and booze and pot and whatever other kinds of drugs are around, usually speed or cocaine.

Right now he's on speed. He stole a good chunk, ground it up in the kitchen with a big stone mortar and pestle, and funneled it into a plastic film canister. (The film inside was used to photograph him pissing and then taking a shower.)

He then snorted a good portion of it, so he probably won't be coming down for another two days.

He's not cold even though it's 27 degrees. He acts as if he doesn't even notice the snow as he walks down the middle of Fifth Avenue, now coated with a couple inches, closed off because of the blackout, reserved for emer-gency vehicles, the sign said. He's not even wearing a coat.

As a worry habit, he rubs the silver robot (a square head with two red eyes and a service door on the lower body section) hanging from his necklace.

He thinks (and says to a non-existent friend) that he's got nothing to worry about, because he's got nothing, but he worries a lot, clenches his teeth when he sleeps, something that Richard (the old pervert) told him he did from watching him sleep.

The thought disgusts Kyler. That guy watching him while he sleeps. It's evil.

Most of what Kyler used to think and believe sort of faded away in the last 30 days. When you're jerking off for the fourth time in 24 hours and it's 3:30 a.m. and you did some dope earlier and you heard from the guy that your naked website is getting 2,700 hits a day and he's making $29.99 off each person who joins, the thoughts about Kyler's ground floor worst-case scenario get more mushy.

He shuts all this off and walks through Washington Square Park at the bottom of Fifth Avenue. He's finally getting cold now, maybe because his brain gave up trying to make sense of his fucked-up life.

WOLF-BOY

Wolf-boy wakes up from a fifteen minute warm slumber. His first thought is that it's only another two days until the New Year's—2003, he thinks. He figures it must be two or three in the morning, even though he doesn't think morning starts until at least eight. He softly gets up and walks over to the window, which overlooks West Seventh Street. His eyes follow big snowflakes from the sky above the trees down to the sidewalk, just as Kyler, the only one on the street, walks by.

Wolf-boy senses something from Kyler, desperation or a faux-confidence that doesn't work, or the idea that he's a used boy, that he's lost and wandering, literally. These are the strange attributes that draw Wolf-boy to other humans. Normal things don't do it for him, alas his current life situation, nearly as bad, but not as evil, as Kyler's.

Wolf-boy slides open the old, broken window, not an easy feat. He scream-whispers down to Kyler. Hark! You there! What's going on?

Kyler sort of slows down, considering that he shouldn't respond to anyone right now, that maybe he's too dangerous or too vulnerable (qualities that go hand-in-hand) to hang out with anyone. Plus, it's been a full thirty days since he's talked to anyone his own age. He's used to being placed on a pedestal, sometimes literally in the living room of the pervert. (It spun slowly as Kyler masturbated in front of the pervert and three of his dinner guests. They ate blood-red venison after.)

Wolf-boy asks, Are you all right?

Kyler looks up to where he thinks the voice might be coming from, up five stories of a three-window wide red brick brownstone. He catches Wolf-boy's gaze from 200 feet away.

Something's going to happen, Wolf-boy knows for sure. He'll be friends with this lonely boy.

Kyler's not so sure. Thirty days of being used like a product dull his senses. Plus, in Iowa, he was used to hanging with the prepsters, the popular kids—the boys who work out and wear Abercrombie, and the girls who do homemade French manicures – those white-tipped fingernails – and they highlight their hair to make it more blond.

Wolf-boy and Kyler stare at each other for ten more seconds, then, out of nowhere, up Seventh Street darts a lone deer, mystical and absurd at the same time, racecar fast and agile, scared, looking at every brownstone wall, every tree, every parked car, for something familiar. Alas, no such sight exists.

Wolf-boy is captivated.

Kyler is startled, maybe even scared. He hides behind a parked van and peeks out over the hood to watch the deer, which both boys can now see is a young buck, a small rack, uneven, one side a miniature of what big bucks have, majestic antlers, on the other just a stick protruding from his light brown head.

The buck almost whinnies and takes some snorting breaths, then seemingly gallops up Seventh Street onto Fifth Avenue, where lucky for the deer, there is not one car driving tonight.

Wolf-boy and Kyler both look around, scanning not only the street and sidewalks, but also all the other windows in the buildings towering over them.

No one is looking. No nosey neighbors, no fussy old ladies, no young professionals waiting for deliveries. Just two people on Seventh Street in the middle of Manhattan on a dark, cold, snowy night.

THE SEX LIVES OF SHADOWS

by Eddie Beverage

I can't really say when my romance with the underworld began. The lack of a time, a place, a beginning and end; it seemed common among the people I met. Our sex lives were dark labyrinths filled with even darker corners and partners we didn't even recognize chained to the walls. No one went there, much less traced his path all the way back to the cellar door. I'd had an urge to fuck anything within spitting distance since I was four years old, so perhaps that's when it all began. I would've fucked my teddy bear if they made them with holes, and why don't they anyway? Then there was that crud-flung night fifteen years later with two sleazy rocker guys in a smoke-choked bathroom on the Sunset Strip and a hot night cap with a teen runaway. Game on.

I came to believe that sex itself meant nothing. It was a cheap thrill, half of which was more imagined than real. In Hollywood, when the lights go out and the rich and famous start screwing, there's a cumslide in the hills and all of Tinseltown bathes in the afterglow. Junkies and prostitutes slurp celebrity jizz from the gutter, impregnating themselves with stillborn fame while star fucker fans scrape leftovers from cracks in the street. Dreams discarded like shitty diapers line the flophouse and when the guys and girls aren't fucking each other, they're fucking cars and money and still can't get close to any of it.

Sex addiction is like any other—you need more the deeper you get. The hard stuff. Fucking some AIDS queen in a West Hollywood alley doesn't cut it anymore. The wolf-like hunger demands more of you. The instinct mutates. I took to roaming the streets of Skid Row at 3 a.m., stalking shadows as they turned corners and disappeared under street lamps, eavesdropping on mumbled voices stuffed with wine bottles, shuffling down Fifth Street past box homes and stalking my prey.

One night in an alley behind the liquor store, I happened upon a drunk, early forties maybe, holding himself up with a wall. His eyes were glassy and couldn't focus and didn't care anyway or maybe they would've mustered the strength to see. I hoped he reeked of liquor and trash. I wanted to smother myself with the stench. Suffocate in it. Die in it. "Hey," I called out. His gaze was estranged from everything. "You and me, partner." His hair was longish and greasy and I wanted to pull it, tear it out, rip it out in bunches while I came inside him.

He staggered over to me. "It's been awhile, hasn't it?" I said. Since he felt anything. Since his pulse had quickened. Since something mattered. The guy didn't say anything. Maybe he's mute, I thought. I grabbed his shirt, spun him around and roughly shoved him against a dumpster. I held a switchblade to his throat. His pants were already loose on his waist and I tugged them down. Skinny, white ass; aren't you going to say anything? He didn't stop me. I slid right in, burying my dick to the hilt. Resist fucker! Resist!! I rammed him against the dumpster a few times, shook him, but he didn't say anything, didn't moan either. I wasn't sure why I even bothered with the knife. Half of them were prostitutes when they had the looks. Taking a fat dick outside their beauty school prime was the best thing that happened to them in years. I couldn't cum so I pulled out. He looked disappointed. "Fuck off," I said. "Fuck off!" He looked more confused than frightened as he wandered off.

A ball of fire rose inside me. The heat. Scalding me. I was filled with dread. Like a junkie fixing with lethal doses who always manages to live. There was nowhere left to go. No greater high. There was no fight left in the ghosts on Skid Row. I had already tried self-imposed sex detox on more than one occasion; lying on my filthy kitchen floor with a pillow over my throbbing prick like some kind of animal. It never worked. I had to go out. To hunt. The air wafting through the window was always thick with the odor of trash and hot dogs; there were no street vendors on my block and I often wondered if the pork smell wasn't a Pavlovian lure by enterprising pimps. It called to me. Get dressed—the underwear with the stretched fuck hole so you can get in and out—fuck that, ditch the underwear and grab the new tennis shoes so you can run if you have to.

Getting into my car, my puny gay Geo, I drove southwest through Koreatown, clenching my teeth and feeling small and insignificant. My blade lay in the passenger seat next to me. It wasn't completely useless. There was that guy from San Francisco who asked me to circumcise him since he was convinced cut queers got more action. Nothing surprises a true sex addict. Even though the foreskin made a great souvenir for a devoted sadist, the blade still craved a real fight. My mind made a random association to a documentary I'd seen about jaguars: solitary creatures, except when it came time to fuck. The male jaguar's penis was covered in sharp backwards-facing spines that ripped into the female and forced her to ovulate. Both sexes had multiple partners. I was fascinated by sex among beasts. Violent, spontaneous boning. Predator on predator. I fantasized about Russian mobsters splashing Vodka on my back, striking a match and gutting me with the bottle. A gangbang with Sudanese rebels. Osama Bin Laden shoving an AK-47 up my ass. Grinning at the far-out images of militant sex, I spun my steering wheel to the left and drove down Vermont Avenue. I didn't need to fly to Afghanistan. I lived half an hour from a war zone.

***

Cops were always mistaking me for a thug. I certainly didn't have a very thuggish name in Marvin. I guess I looked the part though, and I lived in a thug's world. You either packed heat or you felt it. My neighbors in South Central were thieves and killers. Not me. My mama didn't raise no gangsta. But my outsider status didn't end there. I'd never dated a girl. So I wasn't just a good boy but a fag too. Not that anyone knew. I kept it a secret. It was hard sometimes especially around my friend Dvonne who also came from a good family. He had silken, brown skin like a boy king raised in the shadow of the pyramids. Pharaoh's blood; I was sure of it. Long eyelashes and a laid-back style that never faltered.

We used to split the block late at night in my tan Corolla and go surfing: yet another thing thugs didn't do. If we'd gone in the afternoon, folks would've thought homies hangin' ten was a gag. Either that or we would've been beaten to death by skinhead punks. But that was our bond. Salty midnight runs on surfboards we kept hidden under my porch. It wasn't about the waves. We spent most of our time bobbing in the surf under a full moon talking shit about fantasy futures somewhere else and what we'd do if we had a million dollars and when exactly did the Pharcyde stop making good records. The beach was always empty. It was ours.

"You wanna have kids some day?" he'd ask.

"Nah. You?"

"Lotsa work, man."

I'd agree, "No doubt," then chuckle and look away as the tide washed us clear of the subtext that jutted from the surface like sharp rocks.

On weekends we barbecued. Say what you want about the hood, but you can't front on ribs and chicken on a clear Spring day with tendrils of smoke snaking into the air from every yard on the block. My family was tight with Dvonne's. His sister Tamara was the female equivalent of her big brother and his mother was always trying to hook us up. We laughed and joked and no one suspected the battle raging inside of me. It wasn't just I against I, but a war with the macho culture I grew up in. In another life, raised in a white suburb, I would've come out of the closet and a minority of friends and supporters would've rallied around me, but that minority didn't exist on the street where I lived. That didn't mean there were no gays; it just wasn't talked about. My world was an inferno where tolerance was like ashes in the wind. I dated girls like I was supposed to, but they could tell I was different even if my family couldn't, or wouldn't.

I loved Dvonne. I was in love with him. I'm fairly certain of that. But I never did anything about it. What I needed was something to release the pressure. Getting fucked was a hurdle to self-acceptance. I wouldn't be allowed to deny myself after that. It would be real. My sexuality would be irrevocable. No one could take it away from me once I allowed it to happen. And maybe then I could be more frank about my feelings for Dvonne. West Hollywood was the obvious place to make it happen, but I wasn't ready for that. It was too public. It was too done before. And besides that, it wasn't my style. So I trolled the alleyways by the late-night hot spots in my own neighborhood. I was sure there was a cruising scene I'd never witnessed before. A ritualistic underworld pining for a sacrificial lamb.

***

The night's new possibilities made me excited.
Penetrating the ghetto, my cock was so hard I could've
steered my ride with it. Pit bulls strained against leashes
in the front yards of skeletal homes canvassed with
clotheslines and low hanging wires as my dick pressed
against the inside of jeans drizzling pre-cum like slobber
from the mutts' jowls. The only way to do it was to get
out and walk. It felt too safe otherwise. Too much like
some pedophile casing a schoolyard. I wasn't crazy
enough to go after one of the tattooed *chulos* that roamed
the streets in packs. I would have to get one alone.

I parked around the corner from a bar called "The Lair."
The air outside was stiff and combustible. There was an
alley between the bar and a wooden fence with homes in
various stages of decay on the other side. That's where I
went, to the alley, to wait and listen to voices speaking in
Spanish, near or far, I couldn't tell, just background noise
along with the dull thud of bass from a passing car. A couple stumbled out the back door of the bar, black guy,
Latino girl, drunk and kissing and fondling each other.
They didn't notice me and walked off. I stood outside for
almost an hour and there was no one else around, always
voices, but no faces to go with them.

I moved on, cutting across the street to the next block
and standing behind a pool hall where the crack of bil-
liard balls colliding was the only sound. Then I heard
footfalls nearby. It bordered on the absurd how perfectly this guy fit my thug stereotype; black in a wife-beater

top with baggy jeans and even the Timbaland boots. Clutching the switchblade in my pocket with a sweaty palm, I wondered if he had a weapon of his own; a Glock nine millimeter, isn't that what they carried in the hood? The thought made my pulse race. I walked casually in his direction and he sank away. The pursuit led us deeper into the labyrinth of alleys that sliced through the neighborhood. Jaguars with razor dicks bolted through a tripwire jungle in my head.

I rounded a corner and there he was just standing there with his back to me, breathing heavily, almost panting. Moonlight bounced off his chiseled triceps. I remember thinking this guy could hurt me bad if he wanted to. I moved in, holding my blade to his neck with one hand and jamming the other down his pants to grab his hard dick. He grunted a little but didn't fight even though I wished he would have. I unbuckled my pants with a sense of urgency; if anyone saw us we'd both end up shot execution-style right there. There was still no struggle as I ripped his pants down and plowed into him. He was tight as hell. "You like this?" I sprayed in his ear; "You like getting fucked?" He bent over a little and I went deeper . . . tight, tight, tight . . . the friction made me cum. I slapped his ass, smothering it in, and then I shoved him. He didn't say anything. Not a word. But the panting had become something else, nasal, like he was choking. "Wazzup, homie? You don't like the fag steez?" I wanted to fight, a bare-knuckles brawl, then I wanted to fuck some more. As I grabbed his arm and jerked him around to face me, a tear streamed down his face. "Are you fucking crying?" He looked ashamed or relieved, I couldn't tell which. His face was more like a child's than a man, overflowing with emotion as errant light from a streetlamp made rainbows of his tears. I was shocked. It was the most unnatural thing I'd ever seen. Terribly unnatural.

I waved the knife in his face, poking the air with the tip. I craved his fear. But he wasn't scared, far from it, like he'd seen a hundred knives and taken a thousand bullets. Only then did I start wishing a low-riding posse of territorial gangsters might discover us so I could once and for all get some fucking action. When he finally spoke I could barely make out the words as he whispered, "Thank you. Thank you," as if I'd done him the greatest favor in the world.

I turned and ran; I ran hard and I'd never run from anything in my life. I still had the knife clutched in my hand and I must've appeared like a maniac panting and sweating through the maze of alleyways that engulfed me on all sides—I didn't know where I was, yet I couldn't get far enough away. As I stopped to catch my breath, consciousness dimmed then blazed white-hot at the back of my head.

Nothing surprises a true sex addict.

Or so I thought.

HAPPINESS

by L.A. Fields

He laughs in the giddy, I-don't-mean-it-if-you-don't-mean-it way because they're talking about violence again, and it gives him the jitters. Luke sits cross-legged beside him, quiet and serious with his hand cradling the cement burn on his forehead.

"We could, you know," Luke murmurs into the still air of his comfortable room. Luke hates his room, mostly because when he's in it, it means he's been grounded. But Ryan loves it here, the way dirty clothes blanket the floor, making it as soft as a bed and filling the room with the thick, ripe scent of Luke's body. "It's not impossible."

"But it's improbable. I don't think I'd have the balls to do it."

Luke smiles. It's a running joke between them that Ryan doesn't have balls and that Luke must have stolen or otherwise obtained them, since he always seems to have such an excess of nerve. That's how he earned his nasty scrape in the first place, mouthing off to some asshole who had shrewdly picked them out as faggots.

"Seriously picture it," Luke says, his posture unwinding a bit as he relaxes into fantasy. "Blood," he says as Ryan reaches to touch his grated face. "Blood all over the cafeteria, smeared across the tables like catsup on fry day."

"Beautiful," Ryan says, kissing Luke's abraded skin.

"You and I could split up, keep a line of coded communication up through our walkie-talkies. We could track each of those fuckers down."

"And then kill 'em?" Ryan dips his hand into Luke's longish brown hair, lets it feather through his fingers and trickle down his palm.

"No," Luke drawls. "I would shoot out Jake Peterson's knee caps and then let him live. He'll either learn how to run track on his stumps or he can fuckin' rot in a wheelchair."

"You'd really let him live on after us?"

Luke swallows hard and nods. "As long as he hates me. I mean, really just hates me. The way that I hate him."

***

Ryan hates hate. It's a corrosive emotion, kindred of jealousy and anger. It's consumptive and deeply intrusive. It's like love. It gets a stranglehold on your heart, and you feel it against your will. He and Luke discussed the parallel once, lying naked and apprehensive in Ryan's bed.

It started when Ryan couldn't help but ask, "Do you love me?"

Luke swiveled his eyes around to Ryan, his perpetually pot-enlarged pupils like ink blots on the brown parchment of his irises. "I don't know. I mean," he lifted up on his elbows to better explain himself. "Sometimes, I think I do, and it's nice. Like," he touched Ryan's bottom lip, and Ryan gently bit down on the tip. "Like, right now? This is nice. But sometimes I fucking hate that I'm attached to you. That you have control over me."

Ryan smiled. "Doooo as I say," he hypnotizes, his hands up in a puppet-master mime. "Looove me."

Luke smiles weakly and nods. "It bothers me. A lot. Sometimes."

***

At lunch—and they usually spend the lunch hour in the library in the company of nerds and ne'er-do-wells—Luke keeps up a low-toned stream of dialogue. He's been doing "research," studying up on all those school shootings to learn from the mistakes of others.

"Blah blah Klebold, blah blah Jonesboro," Luke says. Ryan's not really listening. He's looking around at the kids doing homework, probably worrying about GPAs and class rank and SATs. Ryan wonders when he stopped assuming that he would graduate from high school.

"Can you picture it?" Luke asks, the question being his launching point for all his gruesome fancy. "Your backpack full of heavy metal, brains sprayed all over a chalkboard. I think I'd sign my name in blood on one of the walls."

Behind Luke's head is a real fake brick wall, and Ryan can picture it: the blood, in cursive, drying from strawberry syrup to brown sticky tack, and Luke's fingerprint at the terminal point of the "e."

"Are you seeing it?" Luke whispers, his voice deep with a dark joy.

Ryan asks, "Would you do your last name too, or just 'Luke'?"

***

Ryan once diagnosed Luke as an artist without talent. Luke has tried drawing (the requisite splatter plots in red ink across the bottom of his math homework), music (clawing at the strings of a secondhand guitar to produce a spine-shattering racket), and writing (short, staccato poems about guns and gore). In all of Luke's endeavors, he never uncovered any untapped potential in himself, no raw ability that only required a little refinement to become legendary.

What he did have was drive, desire, and the suspicion that there was some greatness in him, something worthy of history books. He could just never find it. But under the constant, low-grade torment from the sports heroes, a new suspicion developed: What if he really was worthless?

Ryan was the only person who ever felt the urge to praise or admire Luke. Ryan would horde Luke's sketches, request his songs, memorize his poems. Luke's counselors told him to stop being so morbid. His parents told him to keep the noise down. The kids at school taunted him mercilessly when his poetry notebook was stolen from his backpack and delivered back to him in shreds.

Ryan knew that it meant a great deal to Luke to have another person believe in him, to have someone else reassure him that there was a glory within his nature that "they" just couldn't see.

***

Luke takes to humming "Happiness Is a Warm Gun" everywhere he goes like he's a cursed fallen angel in *Fallen*. His confidence does a roller coaster, escalating slowly and then breaking wildly free. All of a sudden he's telling teachers to go fuck themselves and starting shit in the school's courtyard. He gets bold.

He shoves his stepfather out of his room and locks the door behind him. He turns around and smiles at Ryan, a Cheshire smirk that almost curls up at the ends.

"Let me show you what I got," he says, his eyes a-sparkle. From beneath the cushion of his chair, Luke pulls a bundle wrapped in a towel. He joins Ryan on the bed, sets the bundle in Ryan's lap, and starts to unfold it tenderly.

Lying there is a softly gleaming gun, the kind with a six-bullet chamber (Ryan has no idea what its official title might be). It shines in the pale winter light that filters through the sheet Luke nailed over his window, its black finish like the shiny shell of an insect. Luke hefts it in his palm and strokes the tip of the barrel down Ryan's face. Ryan winces from it.

"Is it loaded?"

"No," Luke says simply, dragging the pointed finger of gun down Ryan's chest.

"Prove it."

Luke takes a second to look indignant, then opens the chamber to reveal six empty holes. Ryan relaxes, reaches out to touch it. Luke hands it over.

"Where did you get it?"

"What do you care? It's not like you want one. Right?" he interrogates.

Ryan fumbles his gaze and it falls to his feet.

"I didn't think so," Luke says, sounding somewhat disappointed. "It's OK. You don't have to . . . do it with me. Just don't get in my way, you know?"

Ryan nods, overwhelmingly relieved and oddly sorry.

"Can we at least fool around with it?" Luke asks softly, circling the barrel loosely with his hand and stroking it with his thumb.

Ryan smiles. "Sure."

***

It all began three years ago when they were freshmen. They met comparing Ryan's crappy home lunch and Luke's suspect cafeteria fare. They bonded over video games, being roughly the same skill level so that their playing was neither boring nor disheartening. They hooked up on a tense night at Ryan's house, a night full of long-lingering glances and unacknowledged elephants.

It happened in the dark as Ryan lay perfectly still in bed and Luke twisted on the floor. Eventually he rolled up on his knees and said, "Um." Ryan reached out and touched the hollow of his neck and in a slow motion instant they were kissing under the soft hum of the air conditioner.

Ryan got up, heart punching, to lock his door, and he and Luke spent the rest of the night in the breathy warmth under the covers, fumbling softly with each other's bodies. Ryan remembers little more than his own highly concentrated panic until they both woke up the next morning, smiling sheepishly, and had a real kiss.

***

The phone rings and Ryan picks it up even though he usually lets the machine get it, since it's usually for his parents and he hates taking messages. His stuttering lack of verbal skills makes him phone-shy.

"Hey," Luke says, his voice electronically distant.

"Hey."

It's quiet for a few beats and Ryan starts to worry. Luke usually calls specifically to talk. He'll snag the cordless phone from his living room and hide out in his room with it, calling Ryan's house at all hours to bitch and rant and vent about Jake Peterson or his stepfather or his math teacher. This silence is something different, something dangerous.

"Hey, um. Don't come to school tomorrow."

Ryan's heart seizes. "I thought it wasn't until April."

"Yeah, well. Something happened. So it's tomorrow."

"What happened?"

Ryan hears a sniff or a shuffle on the other end of the line. "The step-Führer's dead."

"Oh." It gets quiet again.

"Yeah, so . . ." Luke voice is solid, not a tremor or a shake.

"I wanna see you again. Before."

"I don't think so. It might be tough for you. You know, afterwards."

"Like I give a shit. Is it 'cause you have to set something up? I can help you get ready."

"No, Ryan." Ryan blinks a bit at the phone's cradle. He can't remember the last time he was "Ryan" and not "dude" or "man" or something. "I gotta go, OK? I've got shit to do."

"I love you," Ryan says.

Luke is silence until a rushed, "Loveyoutoo," precedes a loud click and a dead line.

NEW YEAR'S EVE 2000

from *Scrappy Soldiers*

by Blair Mastbaum

On this fragile green strip of rocky points and rivers, the native Indian boys emulate gangs, carrying guns and driving souped-up Hondas with rust patches on the fenders. There were four shootings last year, the first murder on the coast in almost ten years.

The once endless forests of Douglas-firs and spruces are now half-destroyed, hacked from the ground for cheap profit and greed, turned into clear-cuts, the most hideous way that land can look, really scary to come upon, like I did once on a hike. I came up over a ridge (whistling!), forgetting my shitty life because of endorphins, and then, there it was: a stream that didn't know where to flow, lost in a sea of fallen pines. It was so truly hideous, so ugly and so evil it took my breath away. Really. I couldn't take in air. I fell to my knees and jacked off right there on the barren mud, somehow thinking that my come would fertilize the soil again and make the giant pines return. My knees were caked with dried brown mud by the time I got home.

***

It's 7:30 on the cold and rainy New Year's Eve of 1999, and Junior is the only person I know to hang out with. He's also the only one I know who has fireworks, which I like to explode on New Year's Eve, a tradition I picked up from my cool hippie grandmother Susan who lived in Hawaii.

Junior has fireworks because he's an Indian and he buys them from the stand on the reservation. We've been planning this night of fiery insanity for months—one of the only things we have left to talk about—and adding to our stash of pyrotechnics mostly from the cash I can squeeze from my mom's purse or steal from the library's donation box, which the old librarian never seems to keep an eye on.

I walk as fast as I can without running to Junior's house, just up the hill, even though it seems worlds away—his mobile home on cement blocks—real Indian-style, he calls it jokingly, but not very funny, just pathetic. The waves sound huge tonight. There must be a storm coming on. At least the weather, powerful and wet, provides some drama around here.

I run the rest of way up the hill toward the woods, pushed along by a steady cold gust off the ocean. When I turn the corner to Junior's mobile home, I see him standing in front of his shack-like garage shed thing in silhouette holding a rocket, backlit by a swinging light bulb on a black cord suspended from the ceiling with duct tape.

When I walk up, he almost smiles, then sighs and sets the huge rocket on an old wooden telephone wire spool.

Why can't some cute boy come gliding down the snowless hill on a snowboard and take me into his arms, and tell me that it'll be OK, that life will become interesting when we escape this deathly quaint little seaside village?

I approach Junior slowly. I'm scared of him when he's not feeling good.

He's punched me as hard as he could several times, once in the eye and I had to lie to my mother about the black eye, saying I got hit in the eye with a golf ball while hiking the woods behind the course.

After punching me, Junior said he didn't know why he did it, or what he was thinking before doing it. One time, he said he didn't even know he hit me, which I totally believed. His expression gives him away always. Right now, he might hit me, so I keep my distance. "Where's the monster?"

"Shut up."

His dad used to burn Junior when he was young with the car lighter, or so he says. He does have a lot of round scars on his dark-complexioned arms.

The monster acts all wise Indian man when he's out fishing with his buddies or picking up firewood or whatever else he does when he leaves the house, which isn't much. He punches Junior and his mom and then cries in the garage.

Monster runs out from the flimsy kitchen door, rocking the whole mobile home as he bounds outside. "Don't touch that rocket!" He ignores me as always. "Those are mine!"

Junior looks at me, sort of for help, sort of watching me watching him get treated like shit and being embarrassed about it. It's totally pathetic. It makes we want to just leave and not have any friends.

Junior and I cower in the corner like we always do when the monster's around. "You smell like shit," he whispers to me.

"I know. I haven't taken a shower all week. Thanks for noticing."

Monster grabs the rocket and takes it inside, acting like a bratty little boy. He screams at Junior's mom right after the screen door slams shut.

She just stands in front of the shitty brown-colored (the only one I've ever seen) kitchen sink washing plastic plates with lukewarm water.

I avoid eye contact with Junior until I'm sure he's shoved the monster and his doormat mom down into his soul, where at least he doesn't have to deal with them right now.

He's going to be a serial killer. It's too bad because he's smart and creative and cool. He paints these beautiful and dark representations of his life—himself mostly, with angry and sad expressions, surrounded by violent boiling oceans, dark pines that swallow up all the sunlight.

I've gotta get out of here. It's too bleak, too depressing to be standing out in the shitty garage of a trailer house with a monster inside on the last day of the twentieth century. "You wanna go out and walk around? I bet there's a party or something. We could snake some beer from someone." I can't ditch him now.

Junior's face is blank. He's shoved his parents' stupidity down and entered this dull hazy place of denial so many times that he forgets how to get back into reality.

He gets lost in the strange, soupy place that he retreats to when he needs to escape. I've occasionally been jealous of this place. I've never been there, I've never seen what it looks like or felt the comfort it provides.

For me, the rain is my comfort. It's reliable and soothing and it smells like life even when life sucks. No matter how many days in a row it rains, I love it even more every time. I like the bluish white color my skin turns after months of clouds and mist. I've grown to despise the sun, maybe because it has forsaken my part of the world.

Junior slowly turns his head and looks at me. His eyes are watery, like he's about to cry—raindrops from inside his head. "I wanna get fucked up," he says, in a way that lets me know he really means it.

I can't tell him not to. I don't have solutions for him. I can only ponder his problems alongside him.

We run away from his house and hop on our wet rusty bikes—stashed out by his mailbox under a moldy blue plastic tarp.

Junior looks back towards the lit-up garage, full of junk, a couple beat up lawn chairs, and an old dented pickup truck. He knows he'll be in trouble when he gets back for leaving, but it's worth it to him. His rebellion is his only power.

I flip off his house without him noticing, and we pedal fast down toward Highway 101, which is just a two-lane road that only gets busy in the summer when the tourists come to stay in the cabins to stare at the ocean like it holds some sort of mystical power, which it doesn't. It's just a huge, enormous, unknowable puddle of cold, salty water with sharp-toothed creatures swimming around in it.

We ride south on Highway 101 toward Cape Perpetua, a 4,500 feet-high mountain of pine trees rising confidently and suddenly out of the ocean. My bike tires thud slightly when I ride over the raised metal reflectors on the solid yellow line down the middle.

We pedal down to where the massive modern cantilevered houses are barely hanging onto the soggy land with stolen redwood and cement supports that drop beyond ocean cliffs. I have no idea how they were built.

We coast along into the loop where the houses' mailboxes are lined up like metal soldiers, all different muted colors and side-by-side in a row meant to look quaint, I think, and we hear someone screaming at us—a deep voice with a raspy edge.

At first, I suspect it's this old man who likes to think that this little loop of road called Albatross Circle is private. His is the only tacky house on the loop, and I think he's ex-Navy. He probably got a settlement from the government for his severe limp. As I start to flip the fascist old man off, I hear:

"Men of the night! What's up, gentlemen?" Someone screams in what I can only call a Frank Sinatra-like voice from the dark shadow underneath a wind-sculpted, tortured and haunted looking old Douglas-fir. The old man would never say gentlemen, let alone men of the night.

We ride back in a circle to see who's calling us as the constant rain picks up.

A power transformer on a tall wood pole explodes above us, causing a bright white flash. It casts the pattern of the tree branches on the grass and lights up the spookily white face of one Ryan Nyquil, a 19-year-old drug addicted, super-contemporary rich kid, famous in local lore even though I've never met him officially. I have hung out in a local coffee shop while he was there once. Him and his black-haired friend, this cute boy who I overheard was visiting from San Francisco, kept going to the bathroom together. Nyquil, the clumsier of the two, came out with white powder on his nostrils a couple of times.

Nyquil's real last name is Nyquist, but from what I know, he has a penchant for over-the-counter drugs, so he was crowned Nyquil, for the drowse-inducing blue gel caps sold in drug stores. He's holding an enormous rocket, bigger than I've ever seen before, bigger than they sell on the reservation, if Junior's not lying.

A kid walks up beside him, zipping his pants back up.

I guess he was pissing in the bushes beside Ryan's huge modern redwood-and-glass house.

I stop my bike in front of Nyquil and the other cool-looking kid, happy to be distracted, happy they want to talk to us, happy that they want to hang with me.

"Where you gentlemen going in such a hurry?" Nyquil asks. He holds his enormous rocket like it's no big deal.

Junior and I stare at it in disbelief. I think it's real dynamite.

Junior steps off his bike and lets it fall to the ground, a subtle "fuck you" to his monster, who stole the bike for him from a Seaside rental place a couple summers ago on his family's only vacation ever, 75 miles up the coast. "Just looking for parties or whatever. What is that thing?"

"Oh, this?" Nyquil holds it up like he's admiring it. "It's TNT, my boys. I secured it over the Internet. The name's Ryan Nyquil. Pleased to meet you." He looks at the cool-looking kid standing beside him.

The kid is tall and skinny, wearing a soaked black T-shirt with the sleeves cut off, showing off his lean mus-cled arms, which I'm sure are goosebumped from the cold wet night. He looks sort of goofy, but sort of amaz-ing, like he's from another planet, like a strange doom-and-gloom character from the planet of Ryan Nyquil with cool light-colored eyes and dark black hair hanging over a greenish-yellow face.

Nyquil pats him on the back, and I'm instantly jealous of his hand on this kid's bony spine. "Explosion.com. My brother informed me. He lives in Israel. You know how they love their ammo." He leans down and lights the fuse.

Junior and I back off behind him instinctually. I watch the kid.

He's not scared at all.

I wonder why he's hanging out with Nyquil, who seems fun but overbearing.

"No worries, boys," Nyquil says as he causally holds

the dynamite, fuse burning, "I know what I'm doing. Directions came with it." He laughs and throws the stick of dynamite into the street.

When it explodes, it produces a bang louder than I've ever heard. My ears feel like they're bleeding. A car's side window, just feet from the massive explosion, blows out and breaks into a million tiny shiny pellets. It blows a crater in the road. Dirt and rocks and pieces of the pavement rain down on the street and the car.

The kid looks at me, precisely and directly, and I know we have to follow him inside.

Nyquil runs down towards his house, hiding in the darkness under the fir tree, and smiles. He holds a mini-flashlight below his chin, horror movie style. "Come downstairs, gentleman. We've got quite a shindig going."

I look at Junior. He doesn't want to stay here. He's scared of chaos from growing up with it. He'd rather be having dinner with some Mormon girl, both parents attentive, with strict and well-defined rules on the wall.

The nameless kid holds his hand out. His eyes shine at me like a bright flashlight, then he looks down quickly. He has a quality about him like stars are always shining on him, but he rejects it and fractures the glow a million different ways.

"I'm Elliot. This is my Indian boy friend, Junior," I say.

Junior gives me a dirty look for saying "Indian."

Nyquil points at the kid. "He's Chaco."

Chaco gives me a look, deep into my eyes. He's got something mysterious inside of him that's drawing me in. It's some strange warmth within, warm and enticing. He nods his head, half-smiling.

"Like the canyon?" I ask.

He nods.

My stomach growls. I watch a drop of rain drip from Chaco's earlobe onto his smooth shoulder. I want to lick it off.

Making sure Junior and Nyquil don't notice, he looks right at me, then turns his head slowly and licks the drop off his shoulder.

It's the sexiest moment I've ever witnessed.

Nyquil holds his arm around Chaco. "He's not from around here. He's up for the winter."

So he's not from here. Fuck. Of course this place, this small, stupid town has no room left for people like him.

"Follow me, boys," Nyquil insists. Nyquil is totally and purposefully weird. He's probably nineteen and way too skinny with bright orange hair (it was once black I figure), veiny forearms, and dirty hands.

I've never known why everyone around here used to talk about him all the time, but now I can sort of understand.

He makes people feel safe in his naïve recklessness. He acts like he knows what he's doing and he's in control of our destiny. Nothing bad will ever happen to Nyquil or me if I'm with him. He's blessed by something—maybe a Northwest spirit that Junior's too abused to know about—and that's nice to be around. It's something I've never felt before.

As we're walking down the lawn, he turns around, pointing at Chaco while he walks. "He doesn't talk much in case you haven't noticed. He says he's deaf, but I don't believe him. I think he's just sick of hearing."

I laugh, sort of, and then stop.

We follow them around Nyquil's modern house and down the stairs along the side, back to his private lair perched on the ocean cliffs and nestled beneath his parents' house. Inside, it's cool and dim like a cave. It feels safe from the explosions and chaos outside, cradled deep underground in moist earth. It's paneled on three walls with one wall of glass facing the ocean, and the floor is carpeted and just dirty enough to feel like you could never ruin it, and clean enough that it doesn't feel like a crack den, not that I've ever been in one.

Three cats lie on the back of a huge gray sectional sofa, and in front of the enormous plate glass, a huge television glares into the room, providing the only light. Junior and I glide through the door.

Two guys that I don't know are slouching on the floor, leaning against the couch, shirtless and skinny, all drugged up out of their minds. One has a black eye in the late stages of healing—green and yellow under pale, thin skin.

The whole scene looks like a fashion ad, but no one realizes that but me. I can't believe this is hidden away in a basement of my small town. I thought I knew everyone and I thought everyone was boring. The new millennium is holding out hope for me.

Nyquil motions to the guys slouching on the floor. "These are the cool cats. Cool cats, this is the Elliot-Man and Junior-Boy." One guy lifts his head up slightly, spilling a cup of hot chocolate that was resting just above his plaid boxers waistband, a mini-marshmallow falls down onto his belt. "Oh." He runs his hand through the brown milky stream filling his belly button and spilling to the floor.

"Sup?" The other guy doesn't bother even looking at us. He stares at the TV. A dog show is on. Strange men and ladies dressed in suits and skirts prance around a soft green floored arena with the exact pace of their perfectly combed dogs, who seem mindless and bored. The camera cuts to the audience, where a little girl, dressed in pink like Jean-Benet, is grabbed, hard, by a man in a police uniform. The camera cuts away quick to an Irish Setter being examined by the judge.

I look at everyone to see if anyone noticed.

Chaco closes his eyes, like he's freaked out by it.

I feel incredibly sober and straight. I take off my raincoat and fall sideways onto an open place on the sectional sofa. "What's up tonight?"

After thirty seconds of pure silence, the nameless guy with the black eye growls, then sort of laughs and throws up into his hands—pasty-looking, pale vomit. It drips through his fingers onto the carpet. A cat jumps up and smells the wet drool spot on the carpet.

I look at Junior, who's still standing awkwardly in full rain gear, looking like a naïve Indian boy, acting as if he still doesn't get the white man after all these years.

He throws me a look like, What the fuck?

Nyquil scolds the vomiting kid. "Don't get that shit on the carpet. Go to the bathroom, man. Fuck-a-loop, dude." He flips on a dim lamp. "Sit, my boy," he says to Junior. "Everything's smooth here." Smooth is his ultimate goal for Nyquil. I can tell by the way he talks. He takes Junior's raincoat and hangs it on a metal hook by the door.

Junior squeezes between me and one of the nameless, shirtless guys and sits on the couch.

"What's your pleasure, sons?" He makes this seem like some sort of fucked-up gentleman's club.

"Just looking for something to do," I admit.

Nyquil asks me, "You in the mood for love?"

Junior sort of ignores him.

"What do you have in mind?" I truly don't know. Is he going to offer me Chaco to take into his room and kiss until my lips are chapped?

"Heroin, the old H, horse, junk, chiba, dope, junkie's delight—the best stuff, too."

I laugh. "You have to be joking." I've never heard of anyone doing heroin except Manhattanites, Mexican gangs, and models. He's putting us on. "Really, what is it, super-strength sleeping pills or some shit?" (Really, I'm just pissed that I never knew any way to get it myself, but I'm not going to admit this to Nyquil.)

"It's from Afghanistan. It's good."

I look at Chaco for proof.

He nods without opening his eyes. He's high, I can tell. His eyes are sleepy-looking and glazed over.

Nyquil reaches into his pocket and pulls out a tiny plastic bag of brownish powder.

My heart races. I wanted to be the first person I knew to do the drugs of jetsetters and models. I thought I was the most sophisticated guy around.

My mom's into culture and fashion and opera and ballet. I thought everyone else's parents were basically dumb, like Junior's.

Heroin should be for cool punk rock bands and international photographers who have Kate Moss on speed dial.

I want heroin now. I need to know what it feels like. "I want it."

"My kind of guy," Nyquil remarks. "Come to my office." He motions me towards this bedroom in the back of the basement. His walls are completely bare except for one photo of Frank Sinatra. "You like my scene?"

What? Heroin and shirtless model-looking boys lying around like bored junkie millionaires? If that's what he means, fuck yes I like it. "It's cool, man. How long have you lived here?"

"Two years. I didn't go to school though. I home-schooled myself. I've seen your pal Junior around. He day-labors for my father."

"Yeah, his dad sucks. I call him the monster."

"My father says he's a good worker."

I don't know what to think. "Oh."

Nyquil reaches for some aluminum foil.

I sit on his messy rich-boy bed, outfitted in shiny gray sheets, as he gets his precious heroin accoutrements from his secret hiding spot, deep in the back of a gunmetal gray roll-top desk.

Nyquil scrapes a small smear of brown tar onto the foil. His space has the vibe of a bomb shelter, supplied and supported by his absent parents' millions. I sort of hate him because he lives the life I've always dreamed of. He's independent. He's a teenage mogul. He's got a car, too. I see the keys—Audi keys—on his nightstand.

Nyquil hands me a tube to inhale the drugs. "Have you done it before?"

I can't lie about this. "No, and I don't have any money. Can I pay you back?"

"No worries." He lights the tar from underneath the foil and it smokes.

I suck in as much as I can, losing some into the atmosphere. That's for the spirits, I think. I would never think about spirits if I weren't doing drugs. I know damn well there's nothing watching over any of us, keeping tabs for our well-being. There may be evil spirits, making surethat each of our lives are pure shit. I'd believe that, but I feel too good suddenly. The heroin takes hold. Everything feels OK. I feel heavy, but light. Happy, but melancholy. Excited, but passive. I take another hit as Nyquil holds his monogrammed lighter underneath the aluminum foil. As I suck in more brown smoke, only slightly burning my throat, I notice Nyquil's personal arsenal—shotguns and a pistol sharing a metal shelf with at least twenty more rockets like the one that exploded the street earlier.

But who gives a shit?

This is what I've been waiting to feel like all my life. Smoothness rushes through my veins. I finally don't have to pretend about this place where everything feels perfect and I feel at ease with the world. I walk back out into the den. Everything's fine with the earth. That's the first time I felt that since I was maybe ten years old.

Junior stares at me, analyzing me in my drugged up state. He isn't happy.

I probably remind him of the monster. I just look at him through my sedation. He's so confused about everything. I see it all now. But fuck it, I think. I'm sure his home life will work out.

He'll figure out something.

All my dreams will come true. I'll be a writer and travel the world for fun. I'll build a house in the country to live in when the world gets too fucked up to handle, and I'll live in a loft with a view of the river. I look at Junior. "You'll be fine, man." I sort of think it, too.

"What?" He has no idea what I'm talking about.

"I mean with your . . . dad and all that shit."

"Yeah, sure." He looks at Nyquil.

"So, Junior-Boy, you want a little piece of heaven?"

What a lame thing to say. He's able to ruin this perfect buzz with words. This buzz and this drug shouldn't be analyzed. It should only be felt.

Junior looks to me for approval, then gets pissed at himself and stands up quickly. "Let's go, bro," he says to me. "My dad's gonna be pissed." He changes his mind fast.

I can't go. I feel too good. Chaco is here. "Fuck your dad. Forget him."

Chaco is staring at me and it feels good.

I want to rest my head on his chest and know that everything's going to be OK. I want to hear his heart beating. I want to feel the warmth of his arm around my neck on a cold rainy night, lying in a bed next to an open window, hearing the raindrops hit a puddle outside that my dog sometimes drinks from. I want to watch his pupils contract when the sun makes its way through the thick gray clouds for half an hour. I want to smell his neck after we run down the stairs to the abandoned beach, the one where that famous shipwreck of 1912 occurred. I want to hike with him into the pine forest and find a beautiful place by a stream, and I want to pull his pants down and pull his shirt off, and I want to lick his chest and smell his neck and . . .

Why can't Chaco sense these thoughts?

I'm still sitting in this room, next to some weird guy, I realize, and I pull my shirt down a bit so no one sees my boner.

Junior sniffs the air and suddenly sits up. "Fuck."

Someone bangs on the sheet-glass window of Nyquil's basement apartment with the million-dollar view of the Pacific.

Bang! Bang! Bang!

Everyone jumps. It's so loud, I feel like the thick sheet of glass could shatter.

Bang! Bang!

The bang is so loud, it reaches into my soul, rattling my nervous system. In his basement, with one huge wall of glass on an ocean bluff, anyone outside can see us, but we can't see them—we can only see our own reflections, bowing in every time the glass is hit.

I look at Chaco.

He's not fazed in the slightest by a scary criminal monster pounding on the glass, bowing it in, distorting our reflections like a fun house mirror. He can't hear it, I realize. He really is deaf.

He finally notices the fear on my face, on Junior's, too.

The banging finally stops, and I see the silhouette of a man standing outside, barely visible through the reflection of me and the shirtless boys, Chaco, and Nyquil on the other side of the sectional sofa. Still watching our reflection, I see Nyquil finally stand up and walk over to the door.

One of the shirtless boys barely raises his head. "Just ignore it, Nyquil."

Nyquil rolls his eyes and opens the narrow glass door. "Careful with the glass, man."

The monster stands there, flexing his chest muscles without realizing it, his arms held out wide like some Sasquatch (no offense, Mr. Quatch). "I installed this glass. I know what it can take and what it can't."

Nyquil backs away and sits on his couch, changing

the channel to the weather. A huge green blob shows days and inches of rain coming onshore.

Junior is frozen; his expression looks like he's just seen a ghost.

The monster sneers at me. I just look away. I have no need for more ignorance, more violence, not when only three hours of 1999 remain.

"Boy, get out here!" he screams at Junior.

Junior grabs his raincoat with one quick stroke and stands beside his dad like a zombie. He should be running the opposite way.

"Mr. Nyquist said he saw you sneaking around his lawn. He figured you were gonna steal his kayaks again. If you cost me this job, boy, you're gonna get it."

Junior doesn't say anything. He looks panicked.

"What do you have to say for yourself? Hanging out with these fairies." He's saying this in front of us on purpose. He and his abused son (monster-in-training) should have left by now. These are not our problems.

I'm so tired of seeing these horrible, abusive scenes I could vomit. And I do.

Vomit fills my mouth before filling my cupped hands. It's hot and smells like acid.

Junior just stands there watching me. I've helped him so many fucking times, and he just stands there as I puke my stomach lining out.

The monster cocks his fist back and punches Junior right in the eye. "Fucking loser, boy. Act straight."

Junior falls back and bumps his head on the alu-

minum railing of the stairs. He hides his face with his
arm and darts out into the rain.

I run to the bathroom through Nyquil's room to
wash my face and hands and mouth out.

Chaco looks at me as I walk into the room. He's the
only one who understands. He looks at me and holds his
hand up like an evil claw.

The other boys are back to normal already, just lying
there, focusing on the TV or the ceiling like it seems they
do a lot.

My oldest friend was abducted by an abusive drunk
monster, and he went willingly. He put up no fight at all.

I'm losing respect for him fast.

When he was twelve, I understood he had to do what
his dad said, but at seventeen, he should be standing up
for himself.

I can't let him get beat up.

But, I want to stay here with the only person whose
made me feel normal in years. I look at Chaco.

He looks back at me.

I hear "go"—well, not hear it, but feel it.

I whisper. Everyone else is ignoring me anyway. "I
don't want to."

Again, I hear him telling me "You have to," deep
inside my head, or chest, or neck, I can't quite figure out
where.

I just stare at him—wisps of hair that over his eyes,
making him look mysterious. I think "why?" and try to
send it to him.

He does nothing, just sits there like nothing is happening, like I'm not racking my brain trying to send him this question. I try again, but nothing. I stand up, doing what I have to do for Junior, the official loser of the century.

Nyquil stands up. "Are you taking off into the night, too, Elliot-Man?"

"I've got some shit to do. You all gonna be around for midnight?"

"Maybe so, my boy. We've got three hours left of the twentieth century. I don't want to waste a minute being bored."

Which I can only guess means doing and giving out as many drugs as possible while exploding bunker-busting dynamite acquired from the West Bank.

It hits me that that's all I'll be doing—wasting the last hours of the millennium on some boy who can't be helped.

As I get to the door, Chaco points at me, barely raising his finger, but you better believe I notice it, even though his hand is in the dark shadowed area between his leg and the couch.

This recognition—even though it's incredibly small—is the best feeling I've ever had in this skinny, abused, little, overly-jerked-off, unwashed body. No one has ever needed me. No one has ever given me more than a second glance, average grades. I'm forever indebted. I can't help but overreact. This is truly a first for me.

MUY SIMPÁTICO

from *Retard Radio*

by Paul Kwiatkowski and John Reposa

I kill the engine and lights . . . ride the clutch and slide the Nova into the driveway. Directly ahead, I can make out the basement lights. As the car glides further toward the garage door, I hear the gravel lightly crunching beneath the tires until the wheels stop. I pull the key out of the ignition and sit still. From virtually every neighboring house, I can hear the same baseball game on in surround sound. I used to hate sports, but now, something about its ambiance I find comforting. Since the grass was recently cut and the bougainvilleas trimmed, there's none of the placid rustling I associate with being home. I find the lack of sound to be slightly unnerving. In its place is a new tone I never noticed; the driveway's floodlights buzz as loud as the ones in the baseball field. They smother the audio contours I've so meticulously notated in my head. My home's parameters seem to have completely changed in the five hours I was gone. I jot a note of these changes in my logbook, then pause outside the car a minute longer, isolating whatever sounds are left before they all bleed into a mess.

A muffled laugh comes from the inside the house. I suspect there's company.

I crouch down before the shrubs so that I can see my so-called boyfriend Dale setting up lights in his photo studio with two new boys we met at the Ultra Mall last month. Both of them are pantless and loving the attention. Dale is debating on posing them in front of a tropical island pull down poster or the standard white sheet backdrop. Rory is squatting in front of a flaccid PJ, a ring pop dangling out of his ass.

I met Dale five years ago; at the time I was sixteen and had run away from home for the last time. I was bored, looking for adventure and romance. I thought it would be cool to blow a stranger through a glory hole at Wendy's. That stranger happened to be Dale. Back than he was a suave guy, a real risk taker who made me feel sexy. He was different from everyone I knew. After he came in my mouth, neither of us could leave our stall. We both knew there was a connection, so I scribbled my number down on a piece of toilet paper and slid it through the hole. We've been together ever since.

Dale made me the star of his dream website project runawayteenfriends.com, but since I turned twenty, his attention has gone to these little underage shits we meet at the mall. I've gone from pinup star to production assistant to Mr. In the Way.

"Come on now Rory, just bite his zipper, he's sweet in the pants. Heh." It's Dale's favorite line and it usually works. He's positioning the boys to allow for the camera's optimum viewing, trying to be goofy and animated, saying weird photographer jokes like, "You're an eagle, so soar. Oh no, you're a dirty little eagle, so spread 'em."

PJ pulls the ring pop out of Rory's ass, slowly pushes

into his mouth, lowers his shades and blows a kiss at the camera. Predictably, Dale goes nuts clicking over forty shots in a span of five seconds. Well, at least I know he's going to be busy for a while. This means I have time for Shiloh and what my old friends have come to call "Retard Radio." Funny, but it seems that with every passing day I have more and more time, all the time. Our dog Sheeba canonically bumps into my leg. Sheeba is a micro version of a greyhound. She's not like other dogs. She's super scrawny and always has this serious look on her face . . . sometimes she just creeps me the fuck out, but all in all she's nice to have around. Right now she wants me to go inside and feed her. She wags her tail in anticipation, black eyes piercing through me; she makes this low-down sound in her throat, a prelude to barking. Not wanting to her to alert Dale to my presence, I decide to give her what she wants.

The house smells acrid with cigarette smoke. Why Dale lets PJ smoke in the house is beyond me . . . OK well, I do know. It's because PJ makes him loads of money . . . but still, so what? Dale can make any punk he wants into a star. Fuck, if it weren't for me and my friends there would be no runawayteenfriends.com.

I pull down the foldaway stairs and climb up to the attic. My Seneca town map is splayed out on a ping-pong table and marked with a bunch of multicolored "Dots" gumdrops. As I unlock my tackle box equipment case, I survey the map, wondering where among the marks and dots Shiloh will transmit from tonight, or if as usual,

he'll keep himself hidden. Anticipation building, I make
my way back downstairs as quietly as I can while carrying my Vietnam-era direction finding field manual, my
radio, my logbook, and my four foot antenna that Dale
would kill me if he saw me stick on top of the Nova.

***

I'm out. Total freedom. I'm driving again. The windows
are down, and the wind blowing through my hair makes
me feel OK. Exiting our street on these expeditions
always gives me a super-rush. My serotonin levels go
through the fuckin' roof; it's like being back to my raver
phase. I tune to the classical music station even though
I don't know any of the composers. I like it because it
makes me feel relaxed and in tune with things. It's a rit-
ual that gets me in the mood. One time I made the mistake of leaving it on that station instead of Dale's usual
classic rock station. He heard it when he used the car in
the morning. For a month, every time he fucked me, he'd
make these over exaggerated composer gestures and go
on and on about what a fag-boy I was becoming. He'd
laugh and laugh about that. Dale can be a real shit sometimes. You'd think I'd be used to the abuse by now . . .
but I'm only getting used to it.

By now the image of these roads has been polished
smooth in my mind. Every curve, every shade, is so familiar. Each day I render the terrain a little better. I often
pull over and take in any nuance of sound that I can
gather for reference later. Like right now, I'm at a red

light and I can hear clinking of a metal fence by the baseball diamond. It's a nice sound. I make a note of its coordinate on my map.

I steer the Nova towards the Super Target and let it take over. It travels this route all the time and knows the way. These moments to myself are really precious but it's hard in this town because everything in the scenery reminds me of something . . . and eventually I start hearing something from the past, or seeing somebody's face. Mostly, I end up thinking about the people I'm most trying to escape from: Dale, my parents, friends. It's hard to get lost in thought in a town where every last thing has been marked up by some past event.

Like right now, I'm passing the liquor store my dad used to buy at. When I was young, I had this big ass Doberman named Major, and when I pass by that liquor store now all I can picture is me and Major sitting in my dad's rusted out Pontiac in 100-degree heat waiting for him to come out of the store and drive us home.

In addition to booze, the store also sells dirty mags, my dad's second passion. He'd end up shopping forever . . . like this one other time I was sitting outside with Major and my friends Marco and Cliff. My dad would pick us all up from school cause he was always unemployed and it gave him a chance to gawk at my female classmates whom he called "up and cummers."

"There's an up and cummer! Huh, huh . . . I'd give her a proper beasting!" he'd say, through gritted teeth.

Totally inappropriate—he loved to embarrass me—

and to make matters worse, Marco and Cliff thought it was super funny and they started saying it too. Anyhow, this one time we were all waiting for my dad to come out and drive us the rest of the way home. Marco had stolen three bottles of Nyquil. We each drank one because we thought it would make us trip. There was no telling how long we'd be there, so we had to pass the time somehow. Both Marco and Cliff got sick in ten minutes flat and threw up thick lumps of half eaten neon green nachos. It was like they never chewed all their food. After twenty minutes, I got bored with watching them vomit, and I was too fucked up to pull Major away from eating it. Luckily, I was able to convince her to pull us back home on my skateboard.

When I got home, Mom was making Ritz cracker breaded chicken casserole. Barely managing to sneak past her, Major and I bee-lined to our hideout spot in the basement below the stairs where Dad kept all his porno magazines in a toolbox. He thought that Mom and I would never dare to look in it. The only reason I knew that he kept them there is because he used to come home all sped-out and say stuff like, "Hey there, little fella! What you think about these little ladies?" while showing me pictures of girls getting fisted and sick shit like that. I never knew how to react. He'd laugh forever like it was funniest thing in the world.

Major was eating wet cardboard on the floor, her ass up in the air, and I realized how much her twat looked like ones in dad's magazines. Something prompted me to take a Philips head screwdriver from the toolbox and stick it inside her cunt. I wasn't really prepared for what her reaction would be, but I definitely did not expect her to bite into my wrist and thrash me around like a cheap toy. Mom came running down to the basement. Totally embarrassed, I told her that Major had attacked me for no reason.

Over the next couple weeks, I convinced my mom to have our vet put Major to sleep.

It was way too awkward having her around.

***

I pull into Super Target and make my way past the seemingly never ending lines of cars to the far end of the parking lot where the concrete splinters and gives way to overgrown weeds. This place used to have a bunch of fast food places daisy-chained together, but all of them went out of business when the Ultra Mall was built. Now it's just a bunch of empty parking lots. I guess the owners of Super Target thought it'd be best to stick it out alone, because they no longer had any competition for the lost people who occasionally come off the Interstate. Anyway, I keep driving through the overgrown lots, eventually getting to a dirt road that used to lead to the old KOA campground that closed in the '80s due to a series of highly publicized campground rapes (which incidentally is how I think my friend Cliff was conceived). I pull in beyond the tree line and kill the engine. I have nicknamed Super Target "Position One" because this is where things tend to start.

I push in a cassette and start recording.

"May 2nd, 7:28 p.m. . . . Audio Notes from Position One," I say.

I switch on the two-way radio. It comes alive with a heavy buzz of undulation. I fight to find the furtive regions beyond the AM signals and truck driver jargon. The night is super active, but nothing I find is suggestive of Shiloh. I close my eyes and wait for any clue in the static.

***

Three years ago, I was in the Super Target parking lot with my friend Marco. We were doing the kind of thing I always did with my friends in high school—hanging around sharing water bottles full of discount vodka, looking for something to eat, trouble to get into. It was noon but all the street lamps were on. Marco is a bad person, but he has a way about him that is unlike the other people in town that really makes him stand out, something that makes you want him on your side; sometimes that is good, but mostly not. He had then and still has a wild ponytail, and his head shaved down the sides, which are forever pinkish, peeling from various degrees of sunburn. His body is lean, angular. At this point, Marco was posing with my old friends Riley, Denise, Cliff, and I on Dale's website, but now he refuses to talk about it. Most of the time we all pretend it never happened when we're around Marco cause he's so nuts. He wears these yellow-tinted, doctor-prescribed sunglasses to "keep his mood up." They're supposed to correspond with his anti-depressants and prevent him from acting out, even though he always does. This night was no exception.

I knew something bad was going to happen when I got there and saw Marco pissing into the coin return of a pay phone. I remember that he was pacing compulsively, smoking a cigarette like a joint. He looked like Major did when Cliff and I mashed up speed and fed it to her. He seemed jazzed about nothing, which always made me worry.

We were in the store getting burritos when it happened. Marco looked out the window and saw a lone boy circling the parking lot pointing a two-way CB radio up at the sky, another two-way radio clipped to his gym shorts, slightly pulling them down.

"Hey, check that goober out," Marco said. He was already drunk, talking too loud. The kid might very well have been in his 20s, but his childishly short shorts, featureless contours, and adolescent facial hair gave him an appearance that was both cherubic and damaged.

"Oh, that's Shiloh from school," I said in a dismissive way. I should've known Marco would find gusto in place of my hesitation.

"You know him?"

"Not really. He used to ride my school bus. I remember this one time he started crying like he was being killed in the middle of class. Teacher had to drag his ass out and send him home," I said.

"Sounds about right, he looks pretty retarded to me. Remember that show *Life Goes On*? He looks like the retarded kid from it. Corky." Already knowing this would not go well. I silently pleaded with Marco to leave him alone.

We stood there watching Shiloh's diffused form move through the foggy parking lot. It was like watching the rare Bigfoot footage that used to play on *Unsolved Mysteries* all the time.

"Fuck this, man. He's pissing me off." I could tell something shady was churning in Marco's head.

I tried to diffuse the situation with some sympathy. "Shiloh's not retarded, he's got autism or something. He's always listening to those CB radios, even if it's straight up static. Rednecks on the school bus always used to fuck with him for it. They'd say he was looking for truckers to blow. Eventually, three of 'em beat him up in the bathroom and curbed his head against the side of a toilet bowl. Last I heard, he was relocated to another school. I haven't seen him in a while."

"What? That's great. Lets go talk to him!" Marco slurred. I only made the prospect of hurting him that much more appealing to him.

"Ahh come on man, forget him. He's probably lost. Besides, he never talks."

"If he can scream, he can definitely talk."

Marco didn't even pay for his beef jerky and Mountain Dew. I paid for both of our food and chased Marco out the chiming door, remembering how much I hate it when Marco fucks with people for no reason.

"Hey, check it out. I bet you anything I can knock that fucker out with one punch," Marco said. My heart dropped and my stomach clenched. I always knew that hanging out with Marco would get me time as an accessory to something.

"No shit, I wouldn't doubt it. Come on, forget him, let's go."

Marco pretended not to hear me. He slowly swaggered up to the boy. Shiloh's thick bifocals obscured any identifiable facial expressions. He had no idea what was happening. I wanted to stop Marco from attacking, but as always, I was scared the fucker would turn on me. So I just stood there. The two-way radio static and buzzing outdoor lighting fused together into a shrill pulse.

"Hey, what's your name? You got a muthafuckin Newport . . . you know, a cigarette? What's with the radios? You lookin' for trucker dick, huh? You too good to talk to me? Come on, talk, bitch."

Shiloh's face resembled someone trying too hard to look at the sun. He never really reacted to Marco.

And then it happened, without any further warning, Marco struck Shiloh in the chest. Shiloh hit the ground instantly. The sound Shiloh's head made on the asphalt was like wet meat thrown down on concrete. I half expected his head to be liquefied.

"Fuck, Marco I don't believe you just did that!"

Shiloh lay there silent and motionless and that was the only thing I could think to say.

Marco looked astonished, not in a guilty way but in an "I'm unbelievably impressed with myself" kind of way. "Puhleeeze, settle down, twinkle toes. Try pulling his shorts down, that should snap him out of it. I bet ya he's faking it."

"Oh come on, let's get out of here." I was thoroughly uncomfortable about collaborating with Marco in this situation but found myself taking on a guilty and conflicting fascination with having some kind of ownership over Shiloh's body.

"Fine, whatever." Marco bent over and positioned himself to rip down the shorts. The drawstring around the shorts was elastic like on children's pajamas. They slid off effortlessly.

Shiloh didn't move. His mouth was frozen into a twisted circle so that his lips curled in above small, squared-off teeth.

Suddenly, I was brought back to second grade, ten years old, letting my Mom's then-boyfriend Noel ease off my underwear. Playing dead just like Shiloh, still not fully understanding why it makes sense as the thing to do.

"Fuck, look how big his nuts are! Shit maybe he is dead . . . holy shit I fucking killed somebody, heh!" Marco said as he poked Shiloh's balls with a stick. I had knelt down and stared into his mouth for a long while. I don't know why, but I felt compelled to stick my index and middle fingers inside, like there would be something for me inside. His tongue was dry, rubbery, and knotted, smaller then I expected it to be. It made sense that he couldn't speak. I searched further down for something to prove he was alive. I ran the top of my index finger over the ridged part above his tongue, it was tacky against my skin but further down his throat my finger became wet. As my knuckle passed his teeth, he jerked forward with a silent heave. I slowly withdrew my fingers as to not arouse suspicion from Marco, content that I knew he was breathing. Luckily, Marco had already lost interest and was throwing beer bottles against a wall. I started walking home, anxious to get out of there before the cops came. I only got a few steps before Marco yelled behind me.

"Hold on, dick milker!"

I turned and Marco threw the CB radio which had been on Shiloh's belt at my head as fast as he could. I just barely caught it.

Marco talks into the other radio so that I can hear his voice transmitted on mine. "Hold onto that . . . this way you can't rat on me. That proves you were here." Then he dropped the radio on Shiloh's crotch.

I clipped the radio on my belt and I walked away, struggling not to be ill.

**

"End of audio notes from Position One, duration of recording: 28 minutes, 15 seconds." Not finding any signal at Position One, I decide to move on.

In transit, I review the list of things I have learned

about Shiloh during the several years I have been listening to him.

1.) Shiloh dislikes confined spaces. The few times I have been able to achieve a definite triangulation of his position from the signals and drove down to observe him, he'd be doing what Marco and I saw him doing when we stole his radio: walking around in circles, almost always in a parking lot or sports field, multiple radios attached to his body, each emitting a different signal. Utterly focused on aligning his body with the structures around him to find the perfect orchestration of signal from each radio. Sometimes this takes hours but he never seems strained by the effort.

2.) Shiloh's family life is really, really, bad. Sometimes during a storm, Shiloh will broadcast from beneath his bed. In reviewing my notes from these instances, I have noticed reoccurring voices present in each broadcast, voices I at first assumed must have been crossed signals. I have concluded that these voices are his family members screaming unintelligible things at each other and breaking things in other rooms of the house. The sounds are always muffled and somewhat distant; at least they seem to be trying to shield Shiloh from the goings-on.

3.) CONCLUSION: Shiloh's dislike of confined spaces combined with his bad family life make him want to escape. It seems that Shiloh creates these Frequency Symphonies in order to distance himself from the people and things around him. Due to his poor eyesight, he

has only to turn on these radios and surround himself
with the perfectly arranged sounds, and suddenly he can
be in a totally different world, completely free of the type
of limitations that holds someone like me back.

Sometimes I think that my attraction is becoming
too abstract in a really unpleasant way. I want to make
contact, force him to be aware of me, but I am not sure
why. Like what difference would it make? I know we're
not going to drive away together in Dale's Nova.
Sometimes I get so pissed cause I think Shiloh is being
fucking selfish, how can he not know I'm out there following his signal . . . fuck.

Deep breath in.

My head hurts.

***

I drive cross-town to Position Two, a.k.a. the Ultra Mall
parking lot. Making my way along the periphery of the
lot, I find my way to the darkest end and activate the
recorder.

"May 2nd, 8:33 p.m. . . . Audio notes from Position
Two." I turn on the radio to a frequency that seems
promising and wait. After a few minutes, I get bored,
open the glove compartment, and pull out a joint. Dale
always keeps the Nova well stocked. I decide to smoke
out while lying on the ground and looking at the stars.
The night is cooler now. This helps soothe the boredom
and reminds me that, while I might have to wait a while
to find these traces of Shiloh in the static, the light from

these stars took millions of years to get to this planet and hit my eye sockets. So, in comparison, thirty or forty minutes of waiting isn't that bad. I listen to the rising and falling signals and wonder where Shiloh is right now . . . why he's not broadcasting.

Predictably, a security guard pulls up and shines a light on me as I lie on the ground. I flip him off and climb back in the Nova. Fucker is watching and waiting for me to leave though. I am just about to fire up the Nova and go when, sure enough, I pick up something promising.

It sounds like Shiloh. The two-way radio crackles and twists until it emits a din of frogs, cicada calls, and finally, footsteps in gravel. I can't say why, but I imagine him being by a swamp, maybe the one that Seneca Canal empties into. Unable to control my urge to make contact, I lift the radio to my mouth and say the first thing that I can think of. "Shiloh? Shiloh? This is Matthew. Remember me? My friend hit you and took your radio. It was a long time ago, but I've been listening to you . . . I have it now and I want to help you, I'm sorry about what we did . . . just . . . I don't know what to do . . . don't know how." Without meaning to, I realize I'm saying this as if reciting a prayer. I guess I've rehearsed it a lot in my head.

The footsteps abruptly stop. After three years, this is my first response. I can picture the utter concentration on Shiloh's face as he stands looking at the radio, surrounded by fireflies hatching from the swamp's murky iridescence. My heartbeat doubles. My bones feel hot. The heat spreads to my chest and head. I might pass out.

I don't know how long I was just sitting here stunned. Here is my chance. I should say something.

My voice creaks, "Don't worry, you don't have to say anything . . . I just wanted you to know that I am listening . . ." After another moment of silence, the footsteps resume and then the signal goes dead. My skin is cold. I see the security guard is still there, so I make a note of the frequency and signal strength in my logbook, fire up the Nova, and head home.

***

When I get home, Nazareth's "Love Hurts" is playing really, really loud and Dale is on the couch getting a hummer by Rory while he and PJ huff helium balloons. I say "hi" and they all laugh at me. Upstairs, I can hear Dale say something about my being his "ass in a glass case" in a squeaky voice. It hurts my feelings, but mostly I'm just relieved he didn't seem to notice all the equipment I am carrying. He doesn't know about Shiloh and I like having at least one secret from him to offset all the times he cheats and lies to me.

I climb the attic stairs and put the antenna and field manual away. I turn on the recorder.

"May 2nd, 9:09 p.m. . . . Audio notes from Position Three." I try to block out the high-pitched giggling of Dale and the boys doing helium balloons downstairs and say this notation a serious voice that'll sound more authoritative if I ever have to refer to these tapes in the future. I turn the radio to the same frequency I'd received signal on at Position Two and wait.

After three more hours of distortion, I'm awoken. My sheets are wet with sweat, tangled around my body. This time the sound churns into something like when two skittish birds flutter inside a thin wire cage. It's a loud brassy sound that must not be coming far from where Shiloh is standing. One by one, a cacophony of radio signals emerge, each one perfectly punctuating the one before it. I shut my eyes again and picture Shiloh as he walks along the periphery of a featureless room turning on several more radios. They crackle with the cross-hatched ciphers of emergency dispatchers layered over truck driver talk and the occasional pirate radio signal. I bring my mouth closer to the radio, about to speak into it, but interrupting Shiloh's performance is a stupid idea. My bare walls redefine themselves under the weight of Shiloh's symphony; heavy waveband downpours stretch the dimensions of my room, insulating the space with a soft pulse. Somewhere, perhaps outside, a car glides down a wet dirt road, starched clothes rustle, a barely audible yawn registers—definitely Shiloh's yawn. . . .

I get up and open the clear plastic wrapper off a just-bought pack of Dots. I lick the bottom of a green one and affix it to the corner of Chicopee Road and Empire Street. Green means that tonight's fix was low accuracy. I pull out my compass, set it to the widest setting, and draw a big green circle around the green Dot—this means that Shiloh could have been anywhere within this large diameter. The more time I spend adjusting the precision of my map, the easier it will be to find a rhythm to Shiloh's habit.

By now, my map looks like a body of water being pelted by raindrops.

Dale thinks its some sort of art project, a.k.a. something to laugh at.

Certain diameters like the baseball diamond, the condemned campground public bathroom, and the area surrounding the swamp are whorled like circular fingerprints. These are the parts I concentrate on. I imagine that inside each tiny vortex, Shiloh is hiding underground.

I pack up the rest of my equipment and go downstairs. Dale must be taking the kids home cause the house has fallen quiet. I lie down, take four Excedrins and picture myself coasting down a two-lane highway. I press on the gas and propel myself out of this life while watching one unfamiliar landscape after another transition and fly past. The image is imbued with that feeling I get sometimes when driving around town in the Nova, the times when I'm able to forget for a moment who I am and disassociate myself from the landscape I'm so used to. It's always such a fleeting thing though, and the next moment I'm realizing that I'm running low on gas and that I have to ask Dale for cash to fill the tank. Something always drags me back to myself. But in this future I'm picturing, this feeling of freedom is different because there's a sense to it that it can go on forever cause anyone I know who could bring it down is like, a thousand miles away. I sit with this thought for a long time and then realize that Dale will, in all likelihood, be home at any second, so I go upstairs and lie back down and hope I can get to sleep by the time he gets back.

I have never been one of those people who dreams of the last pleasant thought they'd had before drifting off to sleep. Although I often wish I knew what that is like.

Tonight though, I'm riding a super rusted-out bike down a hill in the state park. It's all exhilarating and I'm loving it, then pieces of the bike start flying off. First a spring or two, then gears and bits of chain . . . but to my amazement the bike stays upright and in fact keeps going, faster and faster. I try'n squeeze the brakes and they come off in my hands. Next, the front wheel, front axle, and handlebars start wobbling independent of each other. I brace for impact even though I know it's the same old dream I've had dozens of times already.

BLACK 'N' RED: THE PAPER DOLL AND THE CARPENTER

by Nick Hudson

I'm at primary school. I'm a sadistic little shit who, as a baby, looked like a bulbous Chinese girl because of jaundice. My school being a quaint retro-topia and this being the '80s, playtime entails the construction of artworks from dense wooden blocks. I'd laboured over an intricately conceived fortress for a whole lunchtime. War iconography never interested me as much as grandiose architecture and theatrical costumes may have done. I later learned to admire people who could make things out of wood. There we are. Peter, my best friend—inasmuch as we'd put in more hours socially together than we had with anyone else— insistently tugs at me to accompany him somewhere. I steadfastly decline.

Informed by the Jilted Impulse: The citadel meets the Indian kid's foot with vulgar force, scrambling my vision with tears of outrage and inducing blood fever.

Informed by the Architect Undone: The Indian kid's head meets a lengthy hardwood cylinder with diabolical vigour. Inside the theatre of Peter's mind's eye, electrifying palpitations seize his nerve endings, the clarity of his vision landslides into mire, as though his collapse into supine delirium were viewed through Janet Leigh's windshield some way into *Psycho*.

Teacher: What'd you do that for?

Me: Because you do not incur the wrath of the vengeful angel of death.

Teacher: What did Peter ever do to deserve that?

Me: He kicked over my castle because I wouldn't go and play football with the retards.

Teacher: Well, he shouldn't have done that, but you can rebuild it, and violence is not the answer at any rate, you understand?

(Teacher's Internal Monologue: What a spiteful, sadistic, fuckhead of a child. Still, he's only seven and if I moralise piously now, his soul may just be redeemed, so I won't come down too heavy on his little ex-yellow ass. This time. Rice or cous-cous? Shit, that kid's gonna have a serious lump for a few days. Poor cunt. And his daddy built all those hospitals in Calcutta. If I believed in karma, I'd be concerned that a huge Nicholas-shaped typhoon might hit Calcutta and obliterate them, but that'd be totally okay because, by extension, the typhoon'd carry huge chunks of timber over the oceans to this very school, and they'd come careening through the roof and pin Nicholas to the floor by his virulent, screaming head. But I don't, so I don't have to develop a guilt complex for thinking about it. Yeh, in my most neurotic moments, I succumb to a psychosomatic interpretation of the world and get all Catholic about all but the most detergent of thoughts. My chest seizes up.)

Me: Yeh, I understand. But he ruined my castle. And it won't be the same if I rebuild it.

Teacher: Ah, but the next time, it could be even better, huh?

Me (piqued with woe): But, alack, the moment has withered.

Teacher: Don't be such a pretentious, queenie little bitch: you're only seven.

Me: Shit, yeh, sorry. It's gonna take me a good few years to reconcile the idea that, hey, the rigorous revision of my creative work can actually be a wholly constructive enterprise and one whose rewards are immediately and richly manifold. The immersion of myself in the smoothing out and tightening of my prose and the finessing of my metaphorical tools are processes which, while sacrificing the quaint sense of having harnessed unsullied a moment in the butterfly net of my creative purge, are to be undertaken simply because it, well, makes my writing, well, better. Well better. I'm in horrified denial over how long it took to expel from my artistic self the romantic notion that the artist is an unautonomous vessel channelling inspiration through his fingertips, distilling the divine. I'd say it was "unfeasibly puritanical" if the words "lazy" and "pretentious" weren't competing for prominence in my head. Maybe I'm too hard on myself. E-mail me. In the poetic realm however, I've always been ruthlessly surgical. Perhaps the epiphany occurred when I stopped distinguishing between the modes.

**

So I'm getting ahead of myself by a good few intellectual continents. But the point remains, at the time I was savagely fucked off at that whiny little prick for having decimated my grand vision—kick me, not the fruits of my artistic labours, fuckwit—whereas now, well, I try to regard such apparently shitty twists of fate philosophically. But at seven, I had yet to ingest mescaline. Not that mescaline is the zen lube responsible for all of my advanced wisdom and well-reasoned perspective, but in recent weeks it has helped.

That said, according to my bilious, wayward moral geography, I was more in the right to cudgel the fucker than he was to kick over my little wooden castle, and I'd do it again right now and enjoy it.

Haha.

Another afternoon that term: Peter started walking away from me whilst I was talking to him, so my arm lashed out, grabbed his ankle; he landed on his nose, which bled profusely. Once more, I stood righteously in the moral sunshine. I was raised a polite child, and I upheld my parents good intentions in this endeavour, even when it meant breaking a child's nose.

At lunchtime, Matthew compulsively tried to engage me in fights. I was . . . eleven. Perhaps fists and grappling were a crudely manifested expression of his unconscious desire to fuck me, or maybe he genuinely thought I was an annoying, inscrutable little prick. Maybe I fancied him and had my fingers crossed that other undercurrents

were at play. I could never do football. I could never fight.
He had me in a stranglehold. I was gasping. Headlock, my
face pressed into the armpit of his V-neck blue school
jumper. I quite liked the proximity, but breathing was
becoming tortuous, or I would've stayed there, dimly
masquerading a resistance whilst trying to obscure my
hard-on with contorted spasms. I ripped his glasses from
his skull; they fell to the ground. I stamped on them. He
winded me: You stupid fucking faggot. I swear he was
almost crying. Blond guys look pretty cute when they're
on the verge of crying. The actual production of tears
blunders the creamy topography of their features, and if
I were directing their faces as filmed opera, I'd holler
"CUT!" right as the tear ducts moistened. I'm crumpled,
sub-humanoid on the floor getting kicked in the kidneys.
Physically, I like the attention, but as each mercilessly
blunt impact deadens my already scant flesh in that area,
I'm relieved when Hallelujah enters the room:

Teacher: Matthew! Get off him RIGHT now.

Matthew (tumbleweeds of humiliation scurry in a
crossfire with Morricone-bells of righteous indignation,
as the bestial raptors of dizzy myopia circle atop the
scene): He broke my glasses.

Me: He kept trying to fight me. He was gonna break
my arm.

Teacher: Both of you, come with me. You can explain
all of this to the headmistress.

(Teacher's Internal Monologue: I hope Matthew has-
n't seriously injured Nick's back—he does seem to be

bent double. They do seem to fight a lot. I'd rather see Nick win, if I'm honest, which of course, the internal monologue allows me to be; Matthew's been heralded as an academic super-achiever by his parents, and fuck, he knows it. But behind the arrogant sheen he's sterile . . . anodyne. Nick has something . . . rich, untapped complexity . . . He's certainly capable of wry, acute analyses of stuff when you catch him off guard. But . . . achingly . . . he seems always on guard. En garde? I get the impression there's a lot unexpressed by this kid. He never laughs at the same things the others do. He's a bit of a space monkey . . . never participates in straight conversation . . . finds it difficult . . . or boring? I wonder what he does in the evenings at home. Now would I buy tickets to his internal monologue between the ages of eleven and sixteen? Yessir, yes I would. Artaud's silent scream in the (approximate) shape of a boy.)

So we trudge to the headmistress' office. Thankfully I'm no longer hard. Matthew's less weepy. Fucking worm—crying because I broke his glasses and he'd be out of action in maths until the replacement frames arrived.

Maths Teacher: It's OK Matthew, until your glasses are fixed, I can dictate all the problems to you, and you can answer without notation because you're daddy's precocious little supernova and you can work it all out in your head. Suck my cock.

Me: I don't recommend it sir. But he has nice armpits and looks kinda cute when he's on the verge of crying. Quite a broad spectrum of tones in that ostensibly blonde tuft isn't there? Not that I've been studying him in class or anything. Watching his shoulders heave beneath the ill-fitting blazer as he knuckles down to some hardcore problem-solving. Y'know, the chairs in these classrooms really fuck with my back. Little wonder I frequently leave class bent double.

***

Matthew and Joe once stuffed me in a sack, bound my hands and kicked the shit out of me. That was fun. If I could have left the room standing I would've been, uh-huh . . .

When I was six, Christina lunged at my back with a pitchfork because she thought it'd be funny. In retrospect I can empathise with such conjecture. I swore for the first time: arsehole. I feel empty even now at how remarkably un-cathartic it felt. Still, it wasn't as anticlimactic as when I first received a blowjob and my foreskin peeled back raw and inflamed for days because of some chaotically employed teeth. And, at least my first swearword was directed at a girl. I'm daddy's little anarchist. Only this time, I left the scene bent double not because I was hiding anything, but because four pockmarks were sending liquid shards of agony through my body, and even being rude to a girl who liked horses hadn't eased my plight.

Around the age of five I used to read a magazine in which certain pages featured paper figurines designed for the reader to cut out (with the help of a responsible adult, or a parent if the former was unavailable) and dress. These flimsy, 40gsm "dolls" would either be nude and hermaphroditic, or sporting hilarious, wartime undergarments. Their clothes would be printed on the opposite page, and there'd be tabs extending from the arms and legs with which the stylist would attach them to the model free of adhesive, thus allowing for costume changes whenever the social function these flat pixies may have been attending became more informal. The image of these tabs has lingered for some reason. At 13, I remember sitting behind Joe in a German class, visualising for the first time an erotic encounter between myself and another boy. We lie wrapped in and facing each other. He was on top, and binding me into blissful fusion, he'd wrap his legs around my own like thick vines, his arms doing the same until his palms rest on my shoulder blades, squeezing, his fingertips gripping the edges. And how my bones used to protrude—as a teenager I had a body oft-termed "sharp." In the "formal" world I was sitting behind Joe, oblivious to the headmaster's blind abuse of the German language, catatonically immersed in this bruising theatre of bodies melting into each other, flesh invading flesh, skin fluid as liqueur, the absorption of my body into/by another. Wrapped in this vision, the dormant play-mat of my body yielding to the boastful grip of his masterly physique, I had to repeatedly seize the image away from the ruinously intrusive analogy of these paper dolls in tabbed clothes. They were categorically pathetic. In one such aside, I contend that Joe, being the active agent, was the garish floral petticoat pinning the tabs around the shoulders of my vacant, vaguely Nordic-looking doll.

And yet, my body, basically impotent beneath the writhing muscle of the stockier boy, foreshadowed a general, often necrophiliac-baiting passivity in my subsequent erotic fantasies—perhaps the paper doll will be the more enduring image? Who knows? Paper burns so easily. I'll e-mail you. Conversely, I feel at times like I was programmed to desire to be physically worn by other people and discarded like an oily T-shirt only ever used for manual labour on a Sunday. I remained seated for ten minutes after this lesson while things down south cooled off. This time, no amount of crooked posturing was going to shield the evidence of my libidinous wanderings. In the classroom, I performed the illusion of taking notes whilst ineptly sketching a frieze from the above scenario. I realised that my fundamental crappiness in the field of observational drawing rendered all my figure work crudely four-dimensional, which must be a regarded a talent somewhere, by some Svengali of outsider folk art, or a patron of trash . . . a post-ironic tycoon of the terrible.

I decide I'd like to be fucked on ketamine to see if I notice I'm being fucked and to see how great I feel when and if I realise I am being.

***

Instructions to the paper doll:
Follow the analgesic-peddling carpenter down the K-hole and have a hungry, deviant ass torn into the her-

maphroditic landscape of your spotless rear, open wide and have his pre-scored tab grind its seed into the mouth between your hips. Tongue, cock, nails, teeth, and bones lashing paper cuts all over your doleful, dormant pulp, crease at will and have ruddy crayon marks deposited in every crease he leaves in you reaped and gaping. Art Brut.

Instruction to the analgesic-peddling carpenter whose shape is akin to a petticoat, with tabs:

Saturate and disempower the hungry paper doll with your tabs, which are barbed and scarring, like a boy-python squeezing the archaic niceties from, in this case, very willing, if inexpressive prey. He wants to be ribbons, streamers, confetti when you're done brutalising, his body ephemeral and sporadic as lace. He won't feel a thing. Ever again.

On ketamine I would barely flinch; my consciousness a voiceless, hollow ether expanding and contracting somewhere above the scene. Or a string hammock suspended between the barren sandstone fixtures of two opposing rocky outcrops, through which half-thoughts tumble, skittering across the mattress like hail stones infused with incommunicable truths born in some catatonic gulf, but like hailstones, prone to melting before a shape can be discerned. The last pathetic gasp of a fun-fair goldfish, convulsing within the schoolboy's cruel fist before it enters the sea and rediscovers life.

My interior landscape while I'm anally fucked from within the K-hole:

The Overture: A Jungian coyote on a plateau gibbers howled incantations across the decimation witnessed below. A symphony of unintelligible mutterings, the shrieks of feral children, mouths vomiting overly fragrant blossom over wings glued together from the ashes of atrocities by the blood of their memory. My skull, a deep, narrow well, walls jagged with the mosaic of chemical uproar, at the bottom of which dwell my eyes, lolling futilely like grapes in a Halloween cocktail, lidless, disaffected and conspicuously never once wired to my brain or that of anyone else . . . A pearl of insight forms on the lowermost thread of the hammock, with the ebullience of a newborn stalactite.

And the Rest: The son of a recently dead sheepherder perched in solitude and sullenly on a red precipice. Legs rocking, desperately trying to smoke the impossibly tight cigarette he wrestled half-smoked from the stiffening fingers of his dead father. Perhaps hoping to inhale the dwindling essence of his forebears before they attain union with the mescal soul somewhere behind Orion; he spies an altercation between two ranch hands way down in the valley. The valley is empty aside from an eaten old mattress, and a pervasive red light palpates with the rhythm of the air. The puckish onlooker is less interested in the rape-like overtones that are blossoming than in the chest-crippling action of drawing smoke from his dead father's last cigarette and honouring some invented, temporary, cross-dimensional transaction.

It's OK dad. The indignity of having died mid-cigarette need not be advertised.

And from this ledge, it does look a lot like rape. The orphan's breeches are tightening around a vague sensation close to/below his stomach. It hurts. He loosens the breeches and releases his cock. Awareness of the hard-on is negligible next to that of the various wispy smoke spectres escaping his mouth. Perhaps voicing in smoke, spare reimbursements to the dead man who, probably by now, will feel superior in altitude to most living entities—the eagles, and the souls of dead Indian chiefs patrolling the lower atmospheres in shamanic reverie. The bulkier ranch hand, satanically handsome in greasy overalls, has punched the slighter one in the head because he thought it'd be funny; he looks unconscious as his dungarees are shredded in fistfuls by a ravenous pack of knives, his ass crack breached by an angry new weight, but he clearly isn't. Besides, he's conscious enough to recognise that it's not entirely unwelcome. His eyes crinkle at the force of entry however. The cigarette is gone; the dead paper flutters as forgettable as a raindrop into the patterned embellishments on the mattress. The boy starts running his knuckles along his teeth hoping to reintroduce some heat and thus dexterity to his fingers, which if cast in bronze in this position would suffice and furthermore sell as the platonic ideal of cigarette holders. He breathes into cupped hands, the aromatic ricochet is a broth of tobacco and, he aspires, his father's breath. Something is disturbing his colon. That's weird.

Oh yeah. It hurts, so very, very much. Not that I notice.

The moon enters a state of eclipse. The delirious, buckling waif down below—who surely cannot be more than fourteen—bleats garbled protests which the wind's conductor has whipped into semi-muted groans of ecstasy to the ears of the chattering, solar-phallic orphan on the ledge.

Anything goes right? It's an (O)K-trip.

The aggressor, who appears to be about sixteen, has dark, sleek, carpenter's hands, his fringe, a thick anarchic moss strung with rivulets of sweat. A peddler in equine analgesics, he's filling the waif with dreams, pouring his brashness like molasses into virgin intestines. His reasoning is troubled by a whole cocktail of fierce amphetamines; he barely factors his boy as more than a receptacle, an experiment. The boy doesn't necessarily want compassion from whoever fucks him anyway. The aggressor clearly works outside in all weathers, but his skin is so irresistibly immaculate, his eyes are K-holes in themselves. There's something Dorian Grey to his sublime exterior. The waif would consent to anything at the fists and rapiers of this alchemist. His teeth are exposed, lips eked back in grimacing homage to the blasphemous labours underway below. The waif is perhaps a pretender, a countrified faux-naif, diagnosed aged six with a consumptive chest and a fair, unmanly disposition, and he's spent much of his youth indoors baking cakes with his oppressively fussy mother or playing scrabble with his doting but bigoted grandmother. He used to make chains of doll clones from folded sheets of paper, an orgy of connected flesh. So lazy, so smothered, so effete. The carpenter's hands have never been inside oven gloves, or carefully traced with scissors the outline of a crudely drawn doll, but his cunt-starved muscle (remember if you will, this is the patriarchal Old West) has many a time risen inside the oven of a young boy's steaming offal, and he has frequently torn new holes in dolls themselves. The doped kid is unapologetically broken in, thrust by thrust until his insides are numb to his invader's brusque, propulsive enquiry, charred by the sun and bleached of sensation.

One day, the house burned to pieces because his mother never returned to announce that the cakes were amply baked, and since that day he's been an accessory (more a key fob than an accomplice) to his drunk father's shamelessly haphazard attempts at gambling his way into grandiose fortunes. And histories and specifics meld like mucus threads drawn together at the jut of a coyote's mandibles. We converge in . . . an omnibiography?

The carpenter unleashes a valedictory howl from deep within his gut as his energies spasm into the bowl of the waif's rectum. In the auditorium of the orphaned voyeur's head, applause sprinkles the air like confetti at a chemical wedding. Of course now his father's stiff and cold, and there's not even a cigarette from which to draw warmth—an encore unlikely—there's little reason to linger beyond the climax. He stands, hoists aloft his breeches, dusts off with his hands, and surprised at the unusually violent contours of his groin, and the semen splattered across the denim, leaps from the red precipice, a Midwestern Peter Pan soaring through solar halos, his hole a searing, screeching oriole. A kamikaze phoenix blazing towards ground zero again and again.

Carrion raptors trace terrible shapes in the sky. The shapes used to be birthday benedictions wrought in smoke by a swastika-tattooed biplane, back in the day when equine analgesics were used on horses.

He plummets into the chasm, flushed of colour by the kinetic chute of air he now occupies, his shape refracts. He appears of every age at once as though a summation has been attained—robustness compromised, he lands a pinprick, a chime on the pounding head of the drugged, raped waif and melts, assimilated, into a tear, an abstraction, an aborted sentiment. The scream is no longer silent and the profanity has graduated beyond "arsehole," however much this may be implicated in the context of the scream. Atop the precipice a fresh apparition manifests like a coy revolution, ginger as a newborn stalactite and this circular dialogue perpetuates, like the water cycle, a quantity theory of orphaned conceits, or say, a hailstorm of half-truths.

The paper doll emerges from the K-hole raw, saturated with blood and sweat, the latter not exclusively his own. The tears, however, belong to him alone. The carpenter cuddles him close and they dwell in the shallows until the analgesic wears off. The paper doll would like this part to last forever. In every way he is irrigated with light which blinds and heals. Put succinctly, the world is much bigger now, and so is his status within it. He can stand proud and erect. The masquerade has boiled itself dry into redundancy. I guess what I'm trying to say is I'm crying and I hurt all over and I got what I wanted and I just want to stay in his arms indefinitely if you don't mind but I don't know if he genuinely likes me or just wanted to fuck me in every conceivable sense but I don't really care about motives just outcome. When I was nineteen.

LEGEND
by Bennett Madison

Your mother saw that you were covered in answers, so she tried to cover herself in the same. Orchids; vines; a crucifix; a firebird. All those stupid, trashy tattoos of hers—they didn't answer anything. So she pierced herself, hoping to be like you. Her ears jangled when she walked. Then there was the tongue. You noticed the hoops on her nipples through her tank top one day on your summer break, just before she died. It never worked for her. No one could be like you. You didn't have tattoos. You didn't have any piercings or whatever. You were written from the inside. Your face was a Rosetta Stone, flipping constantly with hieroglyphics. Every time I looked at you I found another minor solution.

I know you probably don't remember most of anything. I know you don't have much reason to. I'm just reminding you. Let me remind you what happened, so you can forget again.

***

I learned to tell the future from studying your body when you were asleep, the winter that we lived together, when school was finally over and I read fortunes on the telephone for money. I'd found the job on Craigslist—an ad calling for phone actors. For Entertainment Purposes

Only. The pay wasn't great, but I didn't need to leave the house to do it. And I didn't need to send a résumé. All I had to do was pour myself a glass of wine, light up a cigarette and let the phone start ringing. I was working for Madame Cassandra—the one on TV with the turbans and the muumuus.

Phone psychic. One would have thought . . . but no, it was hard work. Keeping these assholes on the line for an hour, just talking, talking; not even telling them what they wanted to hear, really, but instead, you know, just filling time. I had to stretch every word into five, and keep it suspenseful too, or *click*, and down would go my call average, which meant fewer calls and less money. I got paid by the minute.

I wasn't very good at it, at first, but it turned out that resilient vocal chords and a talent for pointless chatter were basically all I needed to keep people happy. Really, most of them didn't much care what I said at all, just as long as I was saying something. They were not picky about the specifics of their various destinies. They just wanted to hear a voice. If I was having a good day, or was at least drunk, I'd try to throw in a reassuring platitude or a lucky prophecy, just to be a decent guy. Other times I'd predict death, bankruptcy, STDs. If anything, people seemed to prefer the bad news to the good.

It was only occasionally that there would be a caller who really wanted answers. Specifics, predictions, tests of my nonexistent gifts with demands for esoteric knowledge like, "What's my favorite color?" or "If you're

so psychic, tell me my middle name." I actually appreci-
ated those calls, because at least they meant that I was
talking to someone halfway sensible. That was always a
relief. And it was in those cases, late at night, that I
would turn to you, where you slept.

I barely had a grasp on the present. I certainly knew
nothing about the future. Tea leaves, Tarot cards, astrol-
ogy, et cetera. No. What I knew about was you. When
those idiots on the line would ask me things like, "Will I
find love?" or "Is my man cheating?" or "What's my
lucky number?" I would just roll over in bed, phone to
my shoulder, and stare at you and whisper so as not to
wake you—though nothing could ever wake you—I'd
whisper to those strangers: he is breathing steady, lips
open like a baby; he has a wayward, wispy hair on his
shoulder, and he is furrowing his eyebrows a little in a
dream. He is happy.

It worked every time.

"Thank you," they'd all say, and you, in your sleep,
would let out a long, blissful sigh. Question answered.
That was the effect you had on people. Just the very mention of you. The barest description did the trick. It was an
answer that worked so long as I didn't follow its curving
line toward where it became a question all over again.

***

I had loved you the first time I saw you. That was
nothing special: everyone loved you the first time they
saw you. To stand next to you was to momentarily

forego all doubt. In college, where we'd met, you had floated across campus in a haze of contented obliviousness, never knowing quite what you were or where you had come from but always happy in each second for the way the world presented itself to you as a string of courses at an infinite and elaborate meal. "What's this one?" you'd ask, and then eat before waiting for a response.

It had taken you time to turn your attention to me, occupied as you were with your vast realm of possibility. But when you'd finally taken notice, you'd tackled the situation with your usual uncomplicated enthusiasm. You never considered why or how or where anything would lead; you just figured that if something was right there in front of you it must be yours by right.

***

Now college was over, it was winter, and while I busied myself with doing as little work as possible, you had gone out and gotten a job. A shitty job, OK, but one that you actually had to get dressed for, which was more than I could say. You spent your days teaching English to busboys who scrutinized your face trying to learn what they could from it; to old Polish ladies who stared up at you wide-eyed like you were a painting of the Pope; to bleached-out mail order brides who left half-intelligible love notes on your desk after class. Every one of them loved you. And you loved them back. At home, when I tried to talk to you, you answered only in infinitives.

***

Your mother called me for the first time about two weeks after I'd started working the hotline. She was dead by then of course, but that didn't stop her from reaching out, inky fingers through the wire, to find me in your bed and telling the future for ten dollars an hour. She didn't give her name, but I knew it was your mother as soon as I heard the Russian accent. It just came to me. I should have been surprised, but I wasn't. The real surprise was when I realized that I had been expecting her all along. It was Anya. Why shouldn't it have been?

"Who's this?" was the first thing she wanted to know. An easy one. Normally I used a fake name with my customers, but I made an exception for your dead mother.

"This is Sam," I told her. And after a pause, "Is this Anya?"

"Who are you?" she wanted to know again, like she hadn't heard the first time.

A ghost is confused. A ghost is a confusion. A ghost understands just enough to know that nothing is making any sense. A ghost can work the crossword puzzle but not the Daily Jumble.

"Anya, it's me, Sam," I repeated. "Remember—we met at the beach that one time? Alex took me to meet you?"

"The beach," she said.

"The Jersey Shore," I said. "It was two years ago—two and a half."

"I've lost my goddamn page-a-day calendar!" Anya said. "The one with the pictures of the cats!"

It was Anya all right. She'd loved cats. Everything she'd owned had cat on it. You remember the cats, right? You must remember something. "You showed me your tattoos," I reminded her. "I was wearing a blue bathing suit."

It seemed to dawn on her, then, maybe, a little. "Do you like my tattoos?" she asked.

"Yes," I said, even though they were obviously pretty terrible.

"I never liked you much," she said, finally figuring it out. And here I'd thought we'd been getting along. "You reminded me of someone unreliable," Anya muttered. "A real asshole I used to know." There was static on the line. I hung up.

I couldn't make up my mind what was more fucked. The fact that I'd just spoken to a dead woman on the phone, or the fact that it had come so naturally. Ultimately I decided not to think too much about whether it had actually happened at all. Instead, I shuffled to the kitchen in my underwear and topped off my wine. When the phone rang again a few minutes later, it was a woman who wanted to know whether she should move from Warrenton to Leesburg. I told her the cards wanted her to stay put, even though I had no cards and not the first idea where either of those places was.

***

"Your mother called me last night," I said to you the next morning, when you were drying your hair before heading off to teach. I was still half asleep. If I'd really been awake I probably wouldn't have mentioned it.

"My mother?"

"Your mother," I said.

"Anya's dead," you said. And you muttered something under your breath. Something Russian and severe sounding.

For someone with such a way with languages—German, Italian, a little French, even a semester of Arabic—your mastery of your mother's native tongue was limited. The tiny bit of Russian you could speak you'd acquired from years of being screamed at by this crazy, tattooed lady who was part Hell's Angel and part Ukrainian Ricky Ricardo. (The woman who had been dragged across several continents as a girl only to take up with bikers and small-time drug dealers upon her ultimate arrival; who was almost disappointed that she had gotten out before Chernobyl; who seemed unconvinced that she had really made it out at all.)

"Fuck you!" you knew how to shout, in Russian, slanting your eyebrows and waving your hands in an angry halo. "You are not my son!"

"Please kill me!" Anya'd had a tendency to exclaim in her apoplectic moments—which were many. You'd picked that one up too, along with several curses of which you didn't know the exact meaning, except that they sounded serious and ominous, even to someone who didn't know the language at all.

There was one more, I knew. Because, sometimes, early

mornings when I had hours left to doze and you were heading off, my eyes would drift halfway open in bed, and I'd see
you crouched above me—the light from the bedroom window casting a nimbus around you—and you'd be whisper-
ing something low and urgent as you grazed your thumb
back and forth against my forehead while I pretended still
to be asleep. I'll never know exactly what you were saying on
those mornings. I will never know the precise translation,
or how you managed to jam those consonants together in
a way that made them sound gentle. But in the halfway
point between sleep and wakefulness, I knew, without
knowing, that you were saying a benediction.

Then you would go.

***

You were handsome. Your face was smooth and open
from a million years of weather. You were secretly
ancient. But had you learned anything in all those life-
times? Not a lot. That was where your genius was. That
way you had of waking up practically new every morning,
with just the barest memory of yesterday and the day
before and the day before that. Just a dumb, sleepy grin
on your face, eyes bright with all the secrets that you carried—secrets that you yourself didn't even know you had.

***

Your mother kept calling me. She wasn't such a bitch
after the first time, even if she still made no sense.

She said things like, "I'm calling to inquire about the

status of my moose-hide vest." Or, "I would like your accurate prediction as to tomorrow's distraction index."

I started to understand her, after awhile.

"How does he know?" she asked, one time. And without any clarification, I was sure, exactly, of what she meant.

What she was asking was: How could you be so happy? Did you have the secret to it burned into your skin somewhere; some symbol that would unfurl your coiled language of contentedness—this, like, unwavering certitude? If only I could find it. On your chest; on the small of your back; on the tender spot between your balls and the inside of your thigh. I knew it was somewhere. I'd searched for it. It was hidden well.

I didn't tell your mother that. For one thing, she had been dead for awhile. It seemed inappropriate to speak so candidly with a ghost. Instead, I was snappish with her. "I don't know what you mean," I barked at Anya. And then there was static on the line. I hung up.

She didn't care. She kept calling night after night. Sometimes she wanted to know the latest football scores. Other times she wanted to know about you and me. She was hungry for information. I discovered that she enjoyed a good yarn, especially when you were the hero.

***

You'll remember this one; you have to.

A long time ago, there was a snowstorm. There had been a party.

It was January, winter break was about to be over, and you and I had met up in the city, the two of us, the night before we had to be back at school. Remember this? We were nineteen. It was in this snowstorm that I first kissed you.

The snow hadn't started yet when we'd left your Jeep, 9 p.m. on 4th and A. But by early morning, when we returned to the car, the Village was covered, I mean covered, in snow, three inches deep and falling.

You just said, "Let's drive."

"There is a blizzard," I told you.

"We'll be OK," you said from the driver's seat. You were at your tallest and most starry, a sneaky grin on your face. That night, you were bigger than your body.

And I leaned over the gearshift and kissed you. I just did. What else could I do?

I was one thing. And then I was something else. In the space of just one second, a person can become older. But a person is always older anyway, in the end. Even in the middle. One second . . . and then. . . .

***

On the way back to school that morning, 4 a.m., the city opened itself to let us past, fell down flat like a scattering of glittering dominoes. And what I felt most, speeding through a blizzard toward the GW Bridge with you, suddenly someone different at my side, was nothing more than like this shy easing of tension, a gentle slackening that started in my shoulders before moving outward. The next time I would ever have this feeling was almost exactly three years later, with you again in the Egyptian room of a museum.

Anyway. That part comes later. We never would have made it to the Egyptian room at all if Anya hadn't summoned us there, in her own mulish way.

***

She had started calling me almost every night. In life, I had only met her once, but I was starting to feel as if I knew her better than I'd ever known you. I barely saw you anymore those months, if you really thought about it—with you working days and me up until all hours on the telephone. When I did see you, we usually just sat around fighting each other on the PlayStation. You were the more skillful player, but I tended to win.

Meanwhile, my late night conversations with your mother had become almost normal, like talking to an eccentric aunt at the wedding of a distant cousin whom neither of us had ever met. Anya and I talked about Franklin Mint dolls and her tips for breeding Persians. She gave me a truly disgusting sounding recipe for something called American Chop Suey. I read her horoscope aloud to her from the newspaper. She was a Leo, and they didn't get the newspaper where she lived.

***

Most days, I woke up around two. I didn't have a lot to do during the daytime. Leaving the house would have cost money, and pretty much all the money that I had was already allocated to my cigarette and takeout habits. My main hobby, aside from sleeping, was jerking off. Sometimes I'd download porn, other times I'd use the Internet to arrange phone-sex assignations with creepy and much older men in places like Nebraska or Texas. After awhile I started to be concerned that your mother would get confused and call to ask me what I was wearing, or did I have a boner. I had to stop.

I started reading ghost stories. You started staying out later and later every night, going to bars with names like The Dick and The Twat where you danced till your arms were sore even though no one else was ever dancing and smiled at strangers in corners while I stayed home and worked. You were more out of touch every day, unaware of almost everything, but especially of the fact that your dead mother and I were becoming something like friends. You had a purposeful way of overlooking certain key things.

With you gone so much, I would have thought I'd lost my ace-in-the-hole as a psychic, but actually, it barely mattered. I couldn't consult your sleeping figure for answers, but I'd gotten so good at it that I didn't need you anymore. When people asked me to guess their favorite colors, I always said red, and I was almost always right. I started to wonder if your mother was the one feeding me the lines. What would her purpose have been?

One thing I knew from my research was that a ghost doesn't stick around for just no reason. A ghost is looking for something or waiting for something. Some kind of resolution. Maybe she has lost a trinket. A locket or a beloved cameo. Maybe it's something else. In any case, I wanted to help Anya rest in peace. It was the exact sort of thing that I was getting paid for, supposedly.

"Is there something you need?" I asked her once. " Some piece of unfinished business that you're trying to settle? For instance, maybe you're looking for something that you've misplaced."

"I never lost a fucking thing in my life," she said. "I mean, I'm the one who's been misplaced here, OK?"

"Sometimes it's good to take responsibility for things," I said. "For actions, losses. Disappointments. Maybe that's why you're, you know, like, stuck."

"Oh, Jesus Christ," she said, followed by something in Russian. Something ominous and severe sounding.

"Anya," I said.

"Who told you you could call me that?" she spat. And she hung up. I tried not to take it personally—I knew ghosts could be grumpy.

***

One Saturday, you and I spent the afternoon together. It was a little drizzly out, but warm considering the season. We wandered over to the Dog Run in Tompkins Square Park, and paused at the wrought iron pen—you always liked to look at the dogs. That day, you were as sharp as ever, scarf tied just so and tucked carefully into the lapel of your peacoat. Your cheeks were flushed with a rosy, satisfied glow. I thought about saying something to you, but then I couldn't think of what. Anyway, my throat was still sore from talking on the phone for eight hours. So I just reached out and took your hand tentatively, and you brightened, if such a thing was even possible, and kissed me in the crook of my neck.

I tried to remember what it had been like, once, when I had known what to say to you. The way it had come to me, out of nowhere, like it did with my customers now. We'd be sitting next to each other on the subway—or worse, the bus, because you always insisted on taking the bus—or lying in bed, or walking down the block to the bodega, and I'd say something small and stupid, and then you'd say something back, also small and stupid but full of affection and familiarity, and then we'd be talking. Just like that. What had we talked about? I couldn't remember. I wondered what your sign was. I wondered if your favorite color was red.

Of course, I knew that you were a Sagittarius and your favorite colors were hot pink and black. So I turned my attention away from you and back to the dogs careening around the park. It was no use. You couldn't be ignored. You touched your fingers to the dog pen, and as you did it, every one of the dogs stopped its racing and became suddenly mindful of something in their midst. There must have been at least a dozen of them, and all together they looked up, sniffing, eyes darting, and tried to judge the definition of your vast inconsistency. (A galaxy; a monument; a glowing lacuna? All of the above?)

Then, all at the same time, they came to a conclusion. With their owners looking on, huddled together in an opposite corner of the pen, the dogs approached.

There was no barking, just a buzz of excited panting and wagging tongues, tails swinging. The dogs scrambled to be close to you, to press their noses to the gaps in the fence and chew the cuff of your pant leg.

You raised your eyebrows in bewilderment. You reached your hand through the bars and let them taste you, and in a careful flurry of activity, they lined up tidily, as if waiting to receive a foreign dignitary. One by one, every dog licked the palm of your hand, turned, and trotted off. You just shrugged, like, "What the hell was that?"

I knew exactly what the score was. You belonged to the dogs. No. You belonged to the world. You didn't have a mother.

***

Anya was drunk for our final conversation on the psychic line. The only surprise there was that a ghost could get drunk. In life, you'd always told me, she'd been a total lush with a string of sleazy, even more strung-out boyfriends. But who knew they had booze in the netherworld? Did they also have tattoo parlors and meth-head trucker boyfriends? Why not, I guess?

"What's wrong with us?" Anya slurred when I took her call that night. As usual, she hadn't even bothered to say hello. "What did we do wrong?" If I listened hard, I was pretty sure that I could hear the splashing of cheap-o Chardonnay in a tumbler. I had come to be able to recognize the Gallo Brothers by ear. It made me thirsty, so I poured myself another glass from the bottle on the bedside table.

"I don't know about me," I said. "But take a look at yourself, Anya. You're a fucking mess."

"Am I?" she asked. "Am I really?" She was annoyed. "I am a ghost," she said. "You are talking to a dead person. Can you blame a goddamn dead person for being crazy?"

"I guess not," I said. Of course, Anya had been crazy long before the occasion of her death. But I didn't bring that up. "Never mind," I said.

"Never mind is right," your mother said. She sighed scornfully and changed the subject. For the first time ever, she got to a point. "Just tell me—do you love him?"

"Yes," I said, without reservation.

"Does he love you?"

"Yes," I said.

"Just look at him for me. That's all I want," she said.

So I looked at you. She was lucky: you happened to be home that night. You were on your stomach, hands tucked under your pillow. You were snoring happily, chasing down some problem that you would never remember when you woke. It was easy to see why dogs loved you. You dreamt like a dog.

"Does he miss me?" Anya asked.

I paused. It seemed like a bad idea to lie. I didn't want to complicate matters. The haunting was just a novelty

for now, a pleasant break from the truly deadening job of
doling out thoughtless advice for $3.99 a minute, of
which I pocketed around fifteen cents. Ghost or not, with
your mother, it had always been small talk. This was different; deceive a spirit and things have been known to get
out of hand. Rattling doors, flying furniture, shattered
windows and blood trickling in rivulets from the mirror
of the medicine cabinet when you're trying to brush your
teeth. You know what I'm talking about.

Anyway, I thought of Anya as a friend. I owed her the
truth.

"No," I finally said. "I don't think he really does miss
you much."

And that was that. "Thank you for your honesty,"
Anya replied. "That's all I wanted to know."

"You're welcome," I said. But she'd already hung up.

I waited for the phone to ring again with another
customer. It didn't. I checked the dial tone. There was-
n't one.

And you mumbled something I couldn't understand,
and nuzzled into my side, throwing your arm across my
stomach. I wondered: had I been truthful after all? Did
you really love me? I hadn't meant to lie. But when I
thought about it, I had my doubts. Had you ever loved
me, more than you loved, for instance, your green Miu
Miu loafers? Or a bulldog, or an elaborate harmony, or
skiing, or any of the various other things that you were a
zealot for? Or were you secretly just a diplomat, as I'd
sometimes suspected: a brown eyed sleeper agent from a

perkier, more heartless kingdom, sent here to collect and catalogue as many emotions as you could study and approximate before heading on your way with a bounce in your stride?

It was hard to say.

***

Right before Valentine's Day, you asked me, "What did she want?" It had been several weeks since your mother's last phone call. I was starting to think that it had all been a dream.

"Who?"

"My mom," you said. "You said she called you, once. What was her question? Don't you mostly answer questions? Things like that?"

"She didn't have one," I lied. I was perched on the windowsill, smoking a cigarette and ashing onto the fire escape. "She was just trying to be ghostly."

"Oh," you said. "Spooky." And you rose, and walked to me and knelt at my feet and put your head in my lap and cried. It didn't seem unusual in the moment. Small things made you cry all the time, and I'd learned not to take it too seriously. You cried at the end of *Dirty Dancing*; you cried when you stubbed your toe. You cried when I stubbed my toe. Your crying jags were frequent, brief, and sincerely insubstantial. So I put just my arm around your shoulder and pulled you in to myself, and what I wanted to ask you was the same thing I wanted to ask you every time this happened. Who are you, really?

What do you care about, anyway? And like Wendy to
Peter: Boy, why are you crying? But I didn't. It was useless to wonder and even more useless to ask. Your answer
to each of those questions would have been the same. I
don't know; I don't know; I don't know. And then you
would look up, dry-eyed, and kiss me; and when you
pulled away, I would see the corner of your mouth drifting upward. A half-smile, and full.

That night, you fucked me, on my back, on the
kitchen table that was in the so-called living room
because there was no room for a kitchen table in the
kitchen and barely room for one in the so-called living
room for that matter either. You were in me, briefs
around your ankles, feet planted on the creaky wood
floor, rocking, rocking.

"I love you," I said. You pulled hard at my legs.

"I love you," you grunted.

A blank look crossed your face. A bead of sweat
formed on your forehead and made its way down your
smooth chest, down your belly, and disappeared into the
shadow of hair around the base of your cock. Your eyes
were glassy, jaw pushed forward, muscles twisted into the
tight loops of sex. It wasn't really the moment for it, but
I studied your face then, as if for the first time. And what
I saw: in the flash of climax, there was an unspooling—
this dark thread of knowledge that just crawled slow
motion from your half-opened mouth and wrapped itself
around your chest, and your belly, and your cock again
until the whole of you was blackened with its tattoo.

On second thought, no. Not a tattoo. It wasn't ink. You were frozen tall, an obelisk scratched with deep, chiseled markings. Older than the oldest thing, and at the same time, really, the biggest baby. And you remembered. There was the tiniest flicker of recognition behind your pupils, and I could see that, in orgasm, you were spinning momentarily through the caverns of your own name. I wondered if this happened every time. I wondered how I had never noticed before.

"That was amazing," you panted when it was all over and you were back to normal again. You collapsed next to me, prone on the table, and buried your face in my neck.

"Yes," I said.

The next day you told me: "I want to talk to her." There was a nervousness to your affect that I had never seen before. "Where do you think she is?" you asked.

I gave it some thought. I don't know how I knew; but I knew where she would be. It was my job to know those things.

So we went to the museum to look for your mother. The logical place to look would have been with the Romanovs, with Princess Anastasia, but I led the way to King Tut.

And there, on the threshold of the exhibit, you understood what I had guessed as soon as the snow had started to fall on us as we'd climbed the steps outside. It had floated down as we'd begun our ascent, but then, there at the top of the sprawling, granite climb, we'd looked back to discover that the city was blanketed in white, that we could barely see each other through the whipping flakes. What begins in a snowstorm will end in a snowstorm. You hadn't told me that, but you might as well have. It's the type of thing I would have told one of my callers. It makes perfect sense that the universe should have those kinds of laws.

I'd known, then, that you were not coming to find your mother at all. Your mother knew it too.

There was this shy easing of tension, this slackening starting in my shoulders and spreading outward. I was the one she was waiting for. But you, you were going home. How fucked up, I thought, that an end can be so much like a beginning.

"Don't leave me," I said as you outpaced me into the dimness of the Egypt hall. I knew it was too late; you couldn't hear me anymore. And then all I wanted was for you to look back, to look at me one more time. All I wanted was to be written on you, a final mystery and answer etched into the tiniest aspect of your person. Say, to become a greenish freckle on just one of your brown eyes. But your path was already decided for you. You didn't have a word for regret. It was just something baffling and troublesome that blocked you from moving in any direction except straight ahead. And with one hundred and twenty-three ways of saying it, how can an Eskimo ever truly express: What is cold; what is white; what is ancient; what falls from endless gray and then just, I don't know, lingers?

There was a metallic rustling behind me. The clanging of bangles, jewelry, nipple rings, and hoops and ear-cuffs. There in the museum, I could smell the stench of menthols. It was your mother, coming to join me in an exhibit to the abandoned. Her hand grabbed onto my side. Or were we the abandoners? "There you are," she said. Anya's accent had settled, finally, into just a timidity in her inflection. I wasn't surprised to see that her tattoos had finally crawled up her face, where two snakes squirmed on the gaunt ridges of her cheekbones. Her forehead glowed blue with stars. "Have a goddamn Newport," she muttered, offering up a pack.

You didn't hear her. You just marched on, into the glassed-in artifacts, until we couldn't see you anymore. I knew I would never see you again. But I loved you anyway—I mean really loved you. Everyone would always love you. We were all trying to unscramble your biggest riddle. All you could do was smile.

MILES
by Michael Tyrell

After finding Harry's body, Vikram sat at the kitchen table, noting the differences. A spider plant now hung from the ceiling; a bad spackling job had made the walls uneven and patchy-looking. And where was Harry's view of the Empire State Building? Standing up to look through the kitchen window, Vikram saw the new condo, unfinished but conspicuously taller than the other buildings on this residential Brooklyn street.

In March, he and Harry had stood together at the same window, their bare shoulders touching. Dormancy. Safety. The radiator fizzing. Only a car, once in a while, very small, very far away.

"It's inevitable," Harry said. He pointed to what was, at that time, a foundation packed with muddy snow. "They're going to put up an eyesore. This will be the year, I bet."

"Maybe the developers lost their capital," Vikram said hopefully. "You never know. Didn't you say they haven't done anything with it in years?"

Harry sucked his teeth, shook his head. Vikram felt stupid. His penis stirred. Why did embarrassment always make him hard?

Harry made an impatient sign with one hand, as if to say, "Move," but adjusted his position. He slipped behind Vikram, leaning against him. Harry touched Vikram just below his navel. Vikram's penis flipped up against Harry's hand, making them both laugh.

"Is that a protestor I feel?" Harry asked. "Is he picketing against gentrification?"

They jerked each other off right there, in the narrow space between the two countertops, the curtains not drawn, the heater still hissing. The heater's heat and the window's cold glass: a microclimate. Perfect.

Now Harry was dead. Not just a part of Harry, but the entire being, the whole eager and flawed enterprise that had been Harry, was dead. And Vikram could arrange himself (yes, that was the right word, he thought), at this chipped table where he and Harry had eaten at least a dozen times.

His hands weren't shaking; he wasn't hyperventilating; he wasn't cursing himself or luck or God. But though he wanted to close his eyes and keep them shut, he couldn't. He had to keep looking at everything, if only to buy some time. For example, the table's chairs, mismatched from both the table and each other. Cane seating, ripped by the cat. A lead-colored stone lay on the table, next to renewal postcards for *The Atlantic Monthly* and a bent green Post-It with an ornate doodle of a woman's face. Harry had liked to draw people on the subway.

It was all past tense now: Harry's likes and dislikes; Harry's arguments against Vikram's insistence that Vikram keep his homosexuality secret to everyone except Harry; Harry's accusations that Vikram had "used" him, and thus Vikram was a person who was heartless and manipulative, and Vikram was calculating and technical, treating Harry as impersonally as the accounts Vikram handled.

"It's not just that you work for a hedge fund," Harry had said. "Your whole life's a fucking hedge." What did that mean? Barriers around everything, Vikram guessed. They couldn't go anywhere together, not to a museum or a movie, because Vikram was afraid his brother or someone Vikram knew from college would spot them together and identify them as lovers, even if they were not publicly affectionate.

Harry was now lying lifeless on his futon, his hands at his sides, his eyes closed, no apparent cause of death, no rigor mortis, only a refusal to be revived. Even though Vikram was certain he was dead, he still took Harry's handheld shaving mirror from the shower stall and held it to Harry's nostrils. There was no smudge of air. Vikram was seven years old, peeking in on his grandmother's sickroom, when he first observed the mirror trick. He had seen his father leaning over the old woman, holding the glass just under her regal nose. During the weeks she was dying, her almond-colored eyes had seemed to grow lighter in color and less substantial, until Vikram thought they looked like what icebergs must look like, almost invisible. As he watched his father holding the mirror to his grandmother's face, he wondered if maybe a mirror could do something to the eyes—warm them up, maybe, change them until they were once again dark and alive?

When Vikram's grandmother finally died, when his father had finished his business with the mirror, Vikram wasn't allowed to go in and see her body to say goodbye. This was in the London of Vikram's childhood—a city he remembered for the uncles whose accents added a kind of scornful beauty to everything they said.

"Don't hold him back," one uncle said. "He should be allowed to go see her."

But Vikram's father had held him back, and his mother looked down at him, helplessly. Her not doing anything gave consent to this; it was, in its own way, a holding down. Vikram, the sole child among these behemoths, burst into tears and broke free, running out into the back yard.

The subsequent plane ride to JFK was much better, at least initially. Staying awake through the entire transatlantic flight excited him. How could his mother and father sleep? Didn't they want to see how the clouds could become like faces one minute, then change into nothing the next? The dizzying view from the plane window complicated and accelerated his thoughts. He was several months from turning eight years old, but suddenly he felt struck by lightning, full of insight.

To distract himself, he read his book of jokes, most of which weren't very funny, but he still got the hiccups. His grandmother had a cure for hiccups: she'd drop a match into a glass of water and then make him drink the water. How he'd hated it! Knowing he'd never again have to drink that sulfury concoction made him feel happy and guilty at the same time.

When he met up with his older brother Pradip in America, he told him about his sleepless flight, and about the mirror. As soon as Pradip rolled his eyes, Vikram knew he was in for it. "Idiot. The mirror's not a remedy. It's to tell whether they're dead or not. What you should have done on the plane is taken a sip of Dad's Bloody Mary. That would've really put your ass in the clouds."

***

"Sexy Coincidence," the web site was called. There, people posted their "missed connections," opportunities that Vikram knew were usually just the inevitable collisions of urban life, on the subway or the street corner. Not experiences, just illusions, like his old, embarrassingly childish ideas about the mirror and the clouds.

But he'd done it. Bored and lonely, working seventy, eighty hours a week, he'd posted an ad. It was meant for a silver-haired man in a camelhair jacket he'd seen at a listening booth at Virgin Records. The man kept looking at Vikram, who was flipping through DVD slipcases. The man bit his lip once, and Vikram hoped this was a sign: come home with me. The guy was clumsy, too—he kept jotting things down in a notebook, only to drop first the pen, then the notebook. Or was this also a coy signal?

Harry said he liked those words in Vikram's ad: coy signal. Even though he was a far cry from the silvery, sexy klutz, Harry sent Vikram an e-mail. They met up for dinner. There, Vikram told Harry his real name. In the e-mail to Harry, he'd called himself "Miles." That was the name of the first boy Vikram had ever had a crush on, a very hot and very straight épée player from London. They hit it off the first day of Yale freshman orientation, but Miles quickly ditched Vikram for a Goth girl from Florida. Her father was a skydiving instructor, and she hooked Miles up with an offer of free lessons and free accommodations during Christmas break.

"Where's your family from?" Harry had asked at that first dinner. He didn't seem to mind Vikram's lie about the name. "I mean, where in India?"

"The south."

"Kerala?" Cur-rah-la.

A wrong pronunciation, but a lucky guess.

***

Once, Vikram took a chance and met Harry for a bite at Grand Central Station. Big mistake. It was the day Vikram passed his American citizenship test. Did that make him feel a little invulnerable? Maybe. Their meeting in public could hardly qualify as throwing caution to the wind, though. They sat eating pizza in a booth well away from the traffic of evening commuters. Still, Vikram found himself counting how many times he made eye contact with Harry; he even winced when Harry reached across the table and took a sip of his Coke. How obvious was this? Vikram looked at the strangers rushing by, then at the people in the other booths.

Near the shoeshine chairs, a baby-faced musician belted out "Lay, Lady, Lay." He played a few bars, messed up, resumed. People still dropped coins in the guy's cup.

Vikram shook his head. "He's excruciating. I don't care if his cup says it's good for my karma."

"Remember, we're in America," Harry said. "Karma doesn't always work here." Why did he sound so bitter?

Vikram smiled nervously. "We should get going."

"What if someone you knew ran into you now?" Harry asked. "Someone from Yale, or your brother. How would you introduce me?" Harry was smiling cruelly. "Maybe you could use that fake name you told me—Miles, right?"

What a relief when Harry went to the men's room instead of pressing the issue. It didn't matter that Harry was right—Vikram had been thinking of a workable lie, just in case this happened. It was just the unexpected viciousness that appalled him. Harry had told Vikram he'd once dated another Indian guy, also closeted. If he'd tolerated the other guy's situation, why couldn't he do the same with Vikram?

According to Harry, Vikram's predecessor went home to India and married a Bollywood stand-in. His family had financed a start-up Internet company for him.

Vikram wondered how much his own parents had shelled out to hush up the scandal with the Yale boy. They always said they bought Vikram's side of the story, but Vikram knew better. Through money and denial, that mistake had been paid for.

Waiting for Harry to come back from the bathroom,

Vikram tried to focus on the commuters. He kept an eye on one person in the crowd until he or she disappeared from view. Once, he'd read a theory about New York: the landscape is the future, and vice versa. You were drawn to the people you met because unconsciously you remembered catching a glimpse of them years, even decades earlier. He didn't agree with Harry: karma worked everywhere. Maybe this was Vikram's future—sitting and watching, when he should be walking and pretending not to care?

***

One April morning, Harry returned a book Vikram had lent him and a shirt he'd left behind. They'd had a fight the night before.

"So that's it?" Vikram said. "You're dumping me?"

Harry shrugged. "We don't have a relationship. We have a routine." They fought some more.

Yet Harry had not demanded his apartment key back from Vikram. Months later, Vikram text-messaged Harry: Can I see you?

Yes. That message had come in on June 30, 2005, 12:04:56, seven minutes and twenty-nine seconds after Vikram sent his question.

During this second chapter, Harry didn't pressure Vikram about his sexuality. For his part, Vikram credited himself for not repeating his mistakes. He didn't tell Harry he loved him, then retract it, as he'd done the first time around.

That retraction had outraged Harry. "Why did you say it, if you didn't mean it?"

"I never had an occasion to say it before," Vikram said.

"I'm a person, not an occasion."

***

Today, the day of Harry's death, was a Saturday in August. Early in the week, they'd confirmed that Vikram would come over today. When he'd been walking down Harry's block, Vikram had sent the usual message to Harry's phone: On Diamond. On Diamond Street. Meaning, come down and get me. Harry's apartment didn't have a buzzer.

Today, the front door of the apartment building was open, and when Vikram didn't get an answer, he just used the key he'd kept in his wallet.

A suicide. That's what Vikram thought, when he could think again in roughly linear and logical ways and not detonations of recollection, misgiving, horror, and faint relief.

Still, when he rose from the table and made himself go again into Harry's bedroom, his head felt like a wobbly handheld movie camera. Another association saturated with Harry—Harry's selection of disturbing movies: *Rosemary's Baby*, *The Beguiled*, *Children Shouldn't Play with Dead Things*.

Harry was lying on his back, prone. Now I lay me down to sleep.

Vikram looked up at the ceiling, the last thing Harry must have seen. He saw little silhouettes in the light fixture—tiny bug corpses, probably. He flipped the switch, but the light didn't go on.

He looked around for a pill bottle, a razor, a note. Nothing. He leaned over to look on both sides of Harry. Vikram thought maybe Harry had put the evidence underneath his body before he'd died. Would Vikram have to move the body?

Surely, this had been planned. Harry had said, "Come over Saturday." No blood, no wound, body clothed—to Vikram's mind, Harry had gone to great lengths to make his death a mystery. That secret was cocooned in Harry, and only a pathologist could rip it out and say what had happened. And Vikram was sure the body's skin had changed since he'd discovered Harry: Harry's face was becoming chalkier.

Vikram's mind flashed over scenes from crime investigation shows that either had the killer standing over the victim's body or the cops standing over the victim's body. Never this protracted silence and awkwardness—Interior, Day. "Vikram reenters Harry's room and stands over the body. Vikram's face shows only confusion." He'd seen pages from the *Law & Order* audition sides Harry's agent had sent, so he knew what a screenplay of this scene would look like.

The following scene would be worse, of course: disposers and investigators, and Vikram trying to explain that he wasn't involved in this death. He and Harry fucked many times, often with their hands, sometimes otherwise, but what did he really know about Harry? There were text messages between them, but never letters. When the cops questioned him, would they buy the fact that he hadn't killed Harry? Why wouldn't they? The problem was that even if they bought it, Vikram was sure it would get back to his brother, his friends, his parents. They'd ask him how he got in, and he'd have to tell them about the key.

And how would he account for Harry's presence in his life? Harry was thirty-four and he was twenty-four; Harry was a little-known character actor who made guest appearances on *Law & Order*, and Vikram worked at a midtown hedge fund; they had no friends in common and had no connections through aquaintances, work, or school. Harry's being gay was a matter of record among his friends, but Vikram's sexuality was something known only to the dozen or so men Vikram had slept with.

Have fun cowering in the closet the rest of your life.

It must have been Harry who'd sent him that nasty e-mail, back when they broke up, but Harry had sent it unsigned, from an unfamiliar address.

"Closet" wasn't quite it. No, his secrecy was a place like this—this room, whose objects had been deprived of their ownership and function (the television, the light, and the alarm clock all looked absurd now), and where Vikram felt he couldn't move any sooner than Harry could.

Harry was dressed unremarkably: in jeans and a black T-shirt. Under the clothes, there were birthmarks on the chest and ass; Harry's cock, which had a little curve. Once, Vikram had let Harry slip inside him, without a condom.

Vikram decided to leave the apartment and let it play out as if he'd never come. Harry hadn't killed himself for Vikram. Harry had too much pride, Vikram thought. Harry died in some mysterious, sudden way, and it was Vikram's bad luck to arrive just after Harry's death. When Vikram opened the door, a middle-aged man and woman were rattling out of the apartment opposite Harry's.

The woman said something in Polish to Vikram. She smiled. He recognized her from previous visits to Harry's place. She had on big eighties-style sunglasses, the lenses opaque.

"Nice to meet you," Vikram said. Why say something like that? He went back inside Harry's apartment, closed the door. The phone in the kitchen rang, and reflexively he answered it.

"Good afternoon, sir, I'm conducting a survey," the woman said.

"Yes?" Vikram asked.

"I'm doing market research for the MegaMart chain and wondered if you had a few minutes to spare."

"Only a few," he said.

"Thank you, sir." The woman had a thick Western accent. He pictured her at her phone wearing a cowgirl hat and twirling the phone wire like a lasso. He let out a little laugh. Would this show up on Harry's phone record? Would the cops check?

Ever the telemarketer, optimism suppressing any mockery, she got right on with her questions. Would a MegaMart store in his area be a good thing? Would he shop at a MegaMart? Would he recommend the store to a friend?

Yes, he said to the first two questions. I don't know, he said to the third.

"You've been a big help, sir," the woman said. "Could I verify your name and address for our mailing list?"

He gave her Harry's name, address.

Then Vikram went into the only place he hadn't visited since he'd been here. The book room—which was what Harry called it—hadn't changed much. Its bookcases still seemed on the verge of collapse or implosion (so many stacks in so many different directions) and were still the only furniture in the room. A chair would make me stop thinking, Harry had said. I have to keep thinking. Vikram was surprised to hear him say that; he found it hard to believe that actors did any thinking, most of the time.

Vikram stood before the tower of Harry's salt-and-pepper notebooks. Opening one of them, seeing some of the handwriting—would this be the thing to make Vikram lose it?

The red rain of Kerala fell in the year 2004. It was a scarlet coloured rain like blood that scientists later analyzed and came up with surprising results. The "rainfall" consisted of microbial agents that were unlike any others on Earth. This discovery was among the first concrete evidence that life on this planet may have come from an extraterrestrial source, like a bundle of molecules on a meteorite. That would mean we are all aliens and strangers. That would mean Earth maybe wasn't designed for us and that's why it doesn't feel like our home. St. Augustine said this, or was it St. Thomas Aquinas?

Vikram had read this passage a few months ago, while Harry was in the shower. He was still as much as a spy as he'd been when he looked in on his dad checking his dying grandmother. He was inexplicably moved by the British spelling of "coloured." He felt flattered and mystified that Harry had taken the time to look up something about Kerala. Harry must have researched it; Vikram hadn't told him about this.

Vikram had told him that people in his family died young, often in their sixties. They'd been Christians for generations. He had told Harry what he remembered from the Sunday school in Kerala: the minister's poetic description of Judgment Day. For the good—for the people of this church—it would be a moment of rising, a delirious moment like the take-off of that JFK-bound plane Vikram had been on when he was seven years old.

***

Going in for a last look at Harry on the bed, Vikram thought he saw competition.

Marlon Brando looked down from Harry's wall. Vikram didn't recognize this young handsome Brando in his motorcycle outfit. The only movies he'd ever seen with Brando in them were *The Godfather* and *Last Tango in*

Paris. In both, Brando plays characters who have to face death, literally: the dead son heavy with gangland bullets under the mortician's sheet; the dead wife made-up by a Parisian undertaker.

Should he speak to Harry the way Brando spoke to the dead woman—first shouting at her, then begging her to come back?

The proof of Vikram's presence was ineradicable, and it'd just get worse the longer he stayed here, implicating himself, pressing fingerprints on objects, shedding a few hairs when he scratched his head. It was time for him to go. He'd take one of the notebooks. He'd have to have some relic.

Why not take Harry's shaving mirror? He could have that, too. There it was, lying where he'd left it after he checked for signs of breath: on the floor by the bed. He picked it up, turned it over. One side of it magnified objects, and it was in this side he glimpsed his own likeness. Grown up Vikram, behemoth Vikram.

He remembered how, when his grandmother died, he'd run out into the back yard. He'd been sobbing. It was the arguing uncles and parents that made him cry, not grief or the urgent need to see the body. In fact, he was glad he didn't have to reckon with the fact that his father had failed. For all his power and omniscience, his father hadn't used the mirror correctly, and because of that mistake the once-beautiful woman would be lowered into the ground of a country she said she despised.

Would Harry be buried, or cremated? Vikram wanted to know which, and if the first, where.

Even when he was a kid, these details had preoccupied him. It wasn't death that was so complicated; it was the body's afterlife that was complicated. Look at his own family: his grandfather buried in India, his grandmother in the UK, and he, his brother, and his parents would probably end up buried here, in America.

Even now, standing over Harry, it baffled him. He remembered how, on that long-ago flight from New York to London after his grandmother's death, he tried to think forward to Judgment Day—all the people ever born coming back, he and his family included. It was like trying to describe every dream you've ever had in your life, all at once. So many faces! So much distance and confusion! It would be the end of time, and they'd be scattered across the world.

Vikram wondered: Were you alone when you were judged? Or were you with the ones who'd brought you into this world? If there were people besides your family, did you recognize any of the other souls? Were there accidents? Probably not. But at this moment he hoped for them more than he hoped for salvation: accidental meetings and collisions, wrongheaded and sexy coincidences to make him recognize something from his life.

How, how would they be able to find each other when their bodies rose up, on that awful and purifying day?

HAUNTING YOUR HOUSE

by Sam J. Miller

Alan says his apartment's haunted. He got up to pee on his first night there, and the toilet was full of blood. At least it looked and smelled like blood, but the lights were all off and he was scared so he just flushed and went back to bed. Sometimes a woman speaks Chinese and pads quietly through other rooms—so Alan puts a pillow over his head. There are other signs, always in the middle of the night, but he won't turn the light on to investigate.

Me, I find the whole thing fascinating. I'm excited at the thought of ghosts and hauntings. I want to sleep over and meet the Chinese ghost lady and her bloody ghost toilet, but Alan won't let me. He says it's because he doesn't want anything to happen to me, but that's bullshit. He's scared I won't see it, so he'll know he's crazy.

***

I met Alan in the Mott Street Arcade. He was dressed down at the time, wearing a hooded sweatshirt and a pair of blue-green jeans, but they were fancy and fit him too well. I thought to myself: this guy is either some sad grown-up trying to recapture his youth, or some sad john trying to capture a youth. And if the latter, how could he fail to pick me out of that line-up? The other boys in there were pasty-faced nerds or low-level hoodlums. Cute, sometimes, but not hungry. Only I was there for warmth, and possibly for sex. Only I had put mascara on, not that you'd be able to tell in the poor light, and spiked my hair and padded the seat of my pants. Only I had taken the street-trash look and pumped up the volume.

"You're not playing?" he asked me. I was leaning up against *Mortal Kombat* and watching a Japanese boy play *Dance Dance Revolution*.

"No quarters," I said. "I'm so broke it's a joke."

"Really," he said. He hid his smile, and leaned his head towards the exit.

It's sad, but that's what I've been reduced to: leading men on so they think I'm a hustler, when all I am is lonely. Some guys actually walk away when they find out I don't want money. Mutually rewarding gay sex they can get anywhere in New York City: what they're looking for is to flex their financial muscles, pay some straight boy to degrade himself, or to bully them in the most intimate way imaginable.

Walking through his lobby, I thought maybe I'd misjudged him. The place smelled like grease and piss, and puke and cat boxes, and the hundred other gross things you smell in buildings where the maintenance men are permanently out to lunch. So maybe he was just some poor punk rock type after all.

"I know it looks like a total dump," he said, apologetically, when we got to the stairwell. His face was fabulous; features sharp as stone and just as rigid.

I started to say something about how even squalor would be palatial to a kid who sleeps on the subway, but I stopped myself. Better to get to know a guy a little better before you use big words or provide biographical data. The stairs were gross, but evidently not as bad as the elevator we didn't go near. Of course, once we got to his apartment I saw my initial assessment had been right. The man was loaded. His apartment was spacious and stylish, with marble floors and a molded-tin ceiling, and only the faintest whiff of sulfur in the air.

***

"Where'd you pick up a trick like that?" Alan asked me, after a neat combination of tongue and finger action emptied him out all over my skinny chest. His clock said 6 p.m., and the sky in his windows was dark as midnight.

"All my years hustling on the strip come in handy sometimes," I said.

"Shaddup," he said, crashing down next to me. He was on his back, but with one hand he reached over and smeared his sperm into the thicket of chest hair over my breastbone. "You little wiseass." Even though he had no clothes on, his face and body had the same stern blank hardness I'd been so smitten by in the arcade.

"Let's smoke," I said, sitting up, leaning over, picking his nice wool pants up off the floor and rooting through the pockets for his pack of cigarettes. I lit one for him, stuck it in his mouth, and leaned over and lit mine off of his.

"Thanks," he said, and smiled at me, the sort of warm wordless look that once in a while flashed through his stoic shell and kept me coming back. That was my sixth roll in Alan's hay.

"Tell me something, for crying out loud," I said. "What's your earliest memory? How'd you vote in the last election? Your mom's maiden name, your first love, your first fuck."

"I don't vote," he said.

"You're an asshole, you know that?" I said, wiping my chest off with one of his hundred-dollar pillows. He shrugged.

When I first saw him, I figured that underneath the impeccably groomed façade was a total savage, a sick filthy pervert aching to tear me limb from limb and then kiss the bloody pieces. Instead I got a guy in a sexual coma, making me do all the work, so devoid of reaction that the slightest widening of his eyes set me off.

"So?" I asked. "Tell me your life story. Tell me something." I was standing on the bed; I was jumping up and down to underline words.

"What's the matter, Sol?" he asked. "Left your Ritalin at home?"

***

Alan's rent is five thousand a month. Across the hall, down the hall, above him and below him, are Chinese families who have been paying four or five hundred dollars a month for decades. Besides Alan there's one other white guy, a young urban professional on the fifth floor, who also pays way too much.

"I'll let you spend the night some time," he said one afternoon, when he caught me standing in the big empty closet. He'd emptied it out after a bad nightmare—someone was locked inside, and on fire, and when he woke up the apartment stunk like burning hair.

"When?" I asked.

"Sometime. I want to wait, see if things quiet down."

"I don't want to wait until things quiet down. I want to see the ghosts."

"I want them to go away. I need to ignore them. Besides, who knows what they might do to you."

"Why would they want to do something to me?"

***

Alan worked on Wall Street. I think. Something high-stress and high-pay. He told me almost nothing. He said I wouldn't understand, but really I think he didn't want me to know too much. Like I might show up outside Deutsche Bank or UBS Warburg or whatever, drunk and deranged, to destroy his future. Alan's arms, primed by hours at an expensive gym, jerked me from position to position and pinned me down and pulled my hair. My body became a treasury of bruises. In Alan's bed I was a market to be manipulated, a trust fund to be plundered. High school Economics had gone right over my head, but suddenly I understood everything—that money boiled down to violence, that the world is set up for men with money to do what they want.

After coming, Alan passed out. At night he slept badly, what with all the vengeful spirits, so his afternoons were eaten up with long, unplanned naps. The first time he passed out on me, I thought: here is my chance to spend the night. I pictured myself curling up next to him and sleeping soundly on his five-hundred-dollar sheets and seeing the monsters who stalk his bedroom at 3 a.m. But no: his alarm went off at 8 p.m.—he made me tea and sent me out.

***

On account of his apartment being haunted, he did a lot of laundry. Things took on a rotten-fish smell in a matter of hours, even in drawers with sachets of potpourri, or closets with constant incense burning.

Most mornings I'd bum around the Lower East Side, catching breakfast at three different soup kitchens, watching kids in the arcade, and by four o'clock, if the need hit me hard and I had nothing else lined up, and if it was cold enough that I recanted my vow to steer clear of Stony, Stoic, Emotionally Crippled Alan, I'd go to his building and wait for him. It was easy to get inside. The front door was never locked. People on the ground floor were always getting robbed. Landlord wouldn't fix that, either. I'd go down to the basement and wait in the laundry room, knowing that when Alan got home he'd need to do a load of wash. Rich as he was, he could have paid to have everything he owned picked up and washed and cleaned and folded and brought back every single day, but he was ashamed. One time he opened his sock drawer and found a dead octopus, fuzzy with mold. How do you explain that to the cleaners? Alan had a hyper-developed sense of decorum.

November passed away and December replaced it, and I ended up in his laundry room earlier and earlier. By the tenth, I'd just wake up from wherever I'd been sleeping and head straight for the damp heat and the old magazines and the smell of cleaning chemicals.

***

An old Chinese lady walked in, behind her a boy my age, his arms full of bags of clothes. She gave me a big smile, nodded her head. I said hello. It took her forever to put the clothes into the machine, carefully unbunching each piece and laying it down in the round tub. Even across the room I could smell the restaurant stink of those white shirts and pants and aprons—like stale grease and onions, and fourteen hours on your feet.

She moved about the laundry room, telling him what to do, smiling at me from time to time. Yet her helper just stared at me, and the longer he looked, the more I read into his blank cold stare, until it was heavy with all the hate and rage that pulsed in the walls, aimed at me, pulling me down, like gravity. Fifty Chinese households, whose heat and hot water had been turned off long ago by a landlord hell-bent on pushing them out, who had to deal with death threats and bullying and bombs planted in stairwells, saw in my white skin the source of all their problems. I tried to concentrate on my book, some trashy historical romance set in the midst of some other country's genocide, something I found on the curb. He sat on a throbbing washer, watching me.

***

The next day, the same kid was watching Grandma do laundry. At one point, he almost smiled back.

Twice a week we ended up in the laundry room together. Usually he'd play Game Boy.

"I could never get past this level," I said, standing next to him as he played Mega Man 3. "What do you do when you get to the part where the lights go out?"

He showed me.

After that I started sitting next to him, talking a mile a minute like any guilty man. He'd answer my questions, but hardly ever asked any of his own. He was the sort of cute that gets cuter and cuter each time you see it. Most likely he was fifteen, but I don't think it would be statutory rape if we hooked up, because I'm not eighteen myself. Something must be really wrong with me, to keep getting so hung up on these boys with empty faces that could hide contempt or possibly desire.

With my novel's rhythm running through my head, right alongside the storm of pornography that never lets up, I'd think things like:

The machine rumbled away; he came over and stood right in front of where I was sitting. He took the magazine out of my hands and dropped it on the ground so I could see the swelling in his crotch. He unzipped his fly, he unbuckled the belt, he dropped his pants and hung there for a second, showing me, before he clamped his hands over my ears and yanked my head forward. For a half an hour he fucked my face, Grandma folding under-shirts five feet away.

***

Alan never talked to his Chinese neighbors. In the hallway he wouldn't say hi, wouldn't even smile and nod his head. Maybe he thought they were ghosts. Maybe he thought they were all against him, trying to drive him to suicide for some unknown reason, like in a Roman Polanski movie he rented for us once and fell asleep in the middle of.

"I made a friend in your laundry room," I said.

"Oh yeah? That stocks analyst from the fifth floor?"

"No, stupid," I said. "A Chinese kid. He's around my age. Seventeen."

Alan's face didn't change. I'm very vague about my age with him. At first I said eighteen so he wouldn't back off, but sometimes I drop hints that I'm seventeen, hop-ing he'll freak out. He doesn't.

"He speaks English?" Alan asked.

"Perfectly. He was born here."

He nodded like he hadn't thought of that possibility.

"I asked him about your apartment," I said. "He says the family who lived here before you, they got forcibly evicted. There were like twenty people living in this little apartment, the whole extended family."

"Forcibly, how?" Alan asked, his face so white he could have passed for a ghost himself. "Like, the whole family got butchered, burned alive? Bled to death in the toilet?"

"I'm not sure. Whatever it was, it was ugly. Sounds like your landlord is a real asshole."

"Yeah," he said, and shook his head clear. "But he loves me."

"I bet he does. You're paying him an obscene amount of money in rent."

"God, he must hate those people," Alan said. "I can't say I blame him, what with all the money they cost him. He probably would butcher people if he thought he could get away with it."

***

"So, Alan, are you out at work?" I asked once. Maybe it's because he's so straight-laced, and maybe it's because he's so tight-lipped, but I love fucking with Alan. Bringing up all the subjects he hates.

He shrugged. "I try to keep my personal life separate from my work life."

"Meaning . . . you're not out."

"My job is very stressful," he said. "I'm not best friends with my co-workers. We don't talk about our sex lives."

"But they talk about their wives, their girlfriends, no? Just, like, in conversation? As in, my wife loves that movie, or my girlfriend's a vegetarian, so since we moved in together that basically means I'm a vegetarian?"

"Sure," he said slowly.

"Do you talk about your . . . relationships?"

"No."

"Ah." I said.

He kicked my leg, but nowhere near hard enough. "Don't go getting all smug on me, thinking you're this super-liberated man of the world and I'm this sad closet case. It's easy to be free and self-actualized when you have no responsibilities or no ambitions. You try being Mr. Out Fag when you work for a conservative bank that pays you well only as long as you continue to be a certain thing."

"I have ambitions," I said.

"Having ambitions and having the possibility of realizing them are two totally different things," he said.

That was my only visit where anything evenly remotely ghostly happened to me. Alan was napping and I was in his bathroom, taking a long shower, shaving with fifty-dollar shaving cream, applying all sorts of products to my body. My face was caked with apricot-walnut exfoliating scrub, and I was trying my best to knead in a circular motion with the tips of the fingers when I felt hot breath at the back of my neck. I whirled around and nothing was there but a darkened hall, and when I turned back around and looked in the mirror, I saw a woman dressed in white slinking down the hall, away from the bathroom, heading for the room where Alan slept. With all that shit on my face, she must have thought I was him.

I let a couple more weeks go by, and then I stopped going over to Alan's building. One day though, a bleak early-February Thursday where we were set to get tons of snow and then plunge far below zero, I sucked it up and headed for the laundry room. Sure enough, six o'clock came with Alan and a big bag of blood-stained sheets. We put them in the washer and headed upstairs.

"I think I might have to move," he said, when we were done, standing next to the bed with his back to me. "They've started getting into my dreams."

"Really?" I said, more than a little bored.

"Yeah. And they've been leaving messages on my answering machine. And my cell phone voice mail."

"Wow."

"Why do they hate me so much?" he asked.

"No idea," I said, although I had lots of ideas. I stuck a finger into the crack of his heavenly ass, and he whirled around with a terrified look on his face.

Walking out I had something of a swagger, savoring the thought of the sperm he deposited in me and the bite marks on the back of my neck, feeling very oppressed, a vessel for his arrogant white man's libido, a twin for the Vietnamese delivery boy he tipped well and treated with total contempt. An old Chinese couple came in as I went out, the man tall and dignified, and I smiled at him. He responded with a flare of the nostrils, a curl of the lip, seeing me as some sheet-white ogre come to cast him out into the street.

SIXTEEN

by Will Fabro

I'm trying to bum a smoke outside of Puente Hills Mall, almost desperate, skateboarding from one group of ass-hole kids to another. I recognize them from school, and they all give me the same look. I wanna say, Yeah well if you're so cool why the fuck are you hanging out here?

This kid from geometry steps out of Borders and, see-ing me, starts screaming and pointing. "You owe me fuckin' forty dollars, Justin!" I skid to a halt, bleeding everyone's ears. I turn around, pushing off with my foot and whiz away. "Come back asshole!" He starts running.

Over my shoulder I scream, "No fuckin' chance!" and it strikes me as kinda funny. So I laugh, the wind blow-ing into me so hard I can barely see a thing—feels like my eyeballs are being sliced open.

Turning my head back, I see him give up and flip me off. I raise my middle finger in return, and suddenly my body jerks, feet flying off the board. I twist non-acrobat-ically in the air til I tumble to the asphalt, face scraping the hot jagged blacktop. "Ah shit!" I think I hear laugh-ter in the background.

"Real smooth, ace."

I look up, vision blurred. I see the sun blocked by a head, but not really blocked, like the head amplifies the sunlight somehow. Eyes focusing slowly: It's sunshine refracted through blond hair, and he's staring down at me, smirking. He gives me his hand—I think for about five seconds before I reach for it.

He pulls me to my feet with ease. "Fuck off," I mutter, dusting myself off. "What the hell are you doing here anyway?"

Evan's a freshman in college now, over in Chicago. He used to date my sister. They broke up right before they left for different schools.

He folds his arms and looks me up and down, grinning and then grimacing when he gets to my face. "You look like shit." He brings his hand to my raw cheek but I flinch, knocking him away. "Spring break. A bunch of friends from high school are meeting here, and we're all driving down to Mexico for a couple of days." His eyes narrow into mine, he frowns, and I look away. "How's your sister?"

I shrug. "Fuck if I know." I turn back towards him. "Hey, you got a smoke?"

He smiles again and it's getting really annoying. His teeth are so big and his mouth just stretches out his face; I'd forgotten how much I wanted to beat the shit out of him when he smiled. "Aren't you, like, sixteen or something?"

"So fuckin' what?"

He rolls his eyes and reaches into his pocket, tossing me a Camel filter soft pack and a fold of matches. I take two, setting one behind my ear and the other in my mouth, lighting it. It's my first cigarette in a few days, and my head starts to dissolve. The nicotine invades my bloodstream and makes me lose sense, destroying my defenses slowly.

"So hey, what's up?" Evan asks, lighting a cigarette of his own. "How ya doin', anyway?" He looks so relaxed and comfortable, so obviously happy. It ticks me off— our respective stature and posture: He's tall and broad, standing straight up, calm; I'm sort of folding in on myself, my shoulders hunched and my head bowed, like I'm trying to shrink.

"All right I guess." I glance into the parking lot, then bring my head back to him. He stares at me, real serious and pensive. "Hey, you wanna get outta here?"

He brightens at this. "And go where?" A small grin flashes.

"Just drive around. I'm sick of this fuckin' place." I take my eyes off him, look over my shoulder, just to seem detached. I don't know why I feel so desperate all of a sudden.

He turns around and walks towards his car. I watch him, his figure diminishing slowly, until he turns again and cocks his head. I skate over to where he stands.

***

"You know, you really do look like shit," he says as we peel out of the parking lot. "What the fuck happened?"

"You saw," I sigh. "I fuckin' ate it."

He shakes his head. "No, not that. I mean in general. Is everything OK?"

I hate when people who should know better ask a question like that to someone like me. What the fuck am I supposed to say? "What the fuck am I supposed to say? Everything is hunky dory, is that what you want to hear?"

"Whoa!" he yells, laughing a mixture of amusement and frustration. "What the hell is the matter with you?"

I reach for another cigarette and light it, then turn on the radio. A really shitty nu-metal hit from a year ago starts blasting. I keep it on 'cos it's at least something to drown out his voice.

He shuts the radio off and sticks a cassette into the tape deck. That Sonic Youth song "Kool Thing." I can tell from the off-key voice. I really hate Sonic Youth. Evan turns down the volume.

"I'm just worried about you, you seem really different."

"Yeah? How so?" This makes me excited somehow, like, Wow he thinks I've changed, I'm not the same kid he used to know. And it pisses me off that I'm excited—admission of what he must still mean to me.

"I dunno. Like you're really distant. Are you on something?"

"Why? Does that bother you?"

"Answer my question."

A new song starts up, "Safari" by The Breeders. Holy awesome. "Yeah."

Evan shakes his head. "Well it's not like it bothers me, I mean whatever. But you maybe should chill out a little. I mean, you look like a ju—" He catches himself for some reason, shaking his head, sighing. "You look about a thousand times worse than the last time I saw you."

"Why should you care?" I put my feet up on the dashboard and push the seat back a little so that I'm kind of lying down.

I can feel his eyes travel over me. I tense up, fight some kind of urge inside that makes me want to smile.

"Because," he groans quietly, and then he just trails off, the guitars overpowering him. This sweet little-girl/raunchily raspy sexy voice oozes out of the speakers. "He couldn't leave/ Always huggin' the man and cryin' out for me." She sounds innocent and perverse all at once. So cool.

At a red light he looks over at me, his face brimming with what comes off as pity. "Justin, you look horrible."

I dig into my brain for any combination of words that'll form a weapon I can maim him with. "Does this mean you won't fuck me, faggot?" He winces considerably. Success for once.

"Aw, man." He grabs for a cigarette.

"I mean, since you stopped anyway, this probably means there's no way you'll ever want to—"

"Justin, would you shut the fuck up?!" he screams, the cigarette dangling in his mouth, unlit. His hands hold the matches but he can't light them, he's too busy gesticulating as he yells. "Goddammit. Shut the fuck up." He gets quieter as he lights the cigarette, throwing the match out the window.

***

When I was a freshman, Evan started dating my sister
Kate. He'd just gotten his license and he came to pick her
up one night. She was all nervous and changing her
clothes about five thousand times, so when he rang the
doorbell she screamed, "Tell him I'm not ready!"

No one else was home, so I greeted him, "Hey, she's
still doing her makeup or something." He laughed and I
turned around, going back to the living room. He sat
next to me on the couch and watched me play video
games for a while, laughing or cursing or groaning at the
spattering blood and gunshots, until Kate was all zippy
and flushed and horny and ready to go. She had on all
this perfume and really smelled like shit. Evan said, "See
ya later, Justin" when they left.

One weekend our parents had gone on a trip, and
Kate decided to fall in line with some dumb stereotype.
Huge insane legendary party. Like dudes-vomiting-on-
the-front-lawn-and-some-chick-running-around-topless
legendary. So yeah, lame and tame. Our house is pretty
big though, so it was kind of worthwhile.

Everyone was really drunk off of terrible keg beer, and
I was just considered the geeky freshman little brother,
so basically I stayed up in my room and listened to
records and got high. Evan barged in, laughing and reek-
ing of sweat and alcohol.

"Justin, pal! Why aren'tcha partying with us?"

I shrugged. "I don't really feel like it."

"Why not?"

"Everyone seems retarded."

His laugh was boisterous and loud, filling the room and competing with the speakers blasting Black Flag. He crawled—after basically collapsing—onto the bed, patting my head, mussing my hair. "You're a cool kid. You kinda remind me of my brother."

"You have a brother?"

"Yeah. Well, had. He died a few years ago. God, was he nice."

"How did he die?" I watched him, my eyes narrowing to feline slits.

"Drowned." His head swiveled towards me and he flashed a lopsided grin, the kind that looks mixed with a grimace. A tear ran down his cheek and then snot started up. When beautiful people cry in movies, they always manage to look even more beautiful, like the sudden vulnerability and tears enhance their features. When beautiful people cry in real life, their faces look deformed.

My lips were drawn to this disfiguration, his cheek looking like runoff or maybe his face melting. When our flesh touched his eyes flew open, his hand grabbed for my neck, the back of my skull ramming against the wooden headboard.

"The fuck?" he said, shocked. I could barely manage a stammer. Then his lips met mine. As his face pressed into me—so hard it was like our bones were trying to meet—he started unzipping his pants. He tore mine off.

"What are you doing?"

"Shut the fuck up," he muttered.

My fists balled up and slammed against his broad back as he slammed into me. It felt like jabs of a knife, each in quick succession. I thought maybe my guts would explode outta my torso. His hands ringed my neck, pressed against my mouth; I tried to breathe through the gaps between individual fingers. At one point I screamed as loud as I could, idiotically hoping that maybe someone would hear me through the wall-to-wall insanity downstairs.

It was like he went crazy. There was a split second of piercing quiet in between the end of my scream and the beginning of his.

"I said shut the fuck up!"

Fingers curled into a fist that plummeted into my face once, twice, three times, maybe more, I lost count—temporarily lost vision in my left eye. Only 'cos it was swelling.

His hands became fists framing my head, grabbing the pillow hard as his hips accelerated, slamming into my upper thighs. His grunts and the shaking of his body indicated that he was done. I pushed him off, realized I'd come on his T-shirt. It was surprising. The first one not caused by my hand. For a while we just sat there on the bed, refusing to look at each other.

"Oh shit," he said finally, started crying, and I looked at him perplexed, disgusted—everything just seemed strange and off, like a dream on Indian food or Xanax or a nicotine patch.

"Fuck you, Evan."

He looked up, his face leaking and distorted. "I—I'm really, sor . . . sorry," he moaned. It didn't mean it wouldn't happen again, though.

***

Evan was over, supposedly studying for finals with Kate. The beginning of that Sleater-Kinney song was blasting: "I can't find you/ Stay where you are/ I'm in the dark/ Stay where you are." I'd turned the volume down and switched the lights off, pressing my ear to the wall.

I heard her say, "OK so why don't give me three— Evan, I said stop it—why don't—"

"Oh c'mon," he moaned, "who really gives a shit about this stuff?"

"Well, I don't know about you, but I don't really want to lose my 4.0."

"Ugh, God," he groaned. "You're such a fucking tight ass."

"What are you talking about?"

"How long have we been going out and we still haven't—"

"Don't give me any of that, Evan," she hissed. "I really don't—oh yeah, OK great! Just leave!"

"I will!" he yelled as her bedroom door slammed shut. And in less than four seconds my door swung open, Evan bellowing "Your sister is such a fucking cunt!" His figure was visible from the light in the hallway before he slammed the door shut. So dark. He turned on the lights, I squinted at them or maybe him—hard to tell.

"Why?" I asked squeakily, my throat closing in on itself, barely allowing a swallow of nervous saliva.

"Shut the fuck up, I don't need you to talk," he said, voice fluttering over the words, walking over to me as he undid his belt.

"Evan, please—"

His hand flying into my face made a surprisingly crisp sound that ricocheted through the room, flattening me on my back. I saw tiny specks of multiple stars dancing.

Before he plunged in, he covered my mouth, muffling any noise I could make. It was quiet then, aside from the crickets chirping outside and his panting in my ear.

When he left I turned the record back up. A scream over a glorious cacophony of guitars and the words "Heart attack/ Hit and run," and then the record stopped.

***

At one point he began forgetting to stop by my room whenever he was over. Kate must've started putting out. I'd always clench up in bed whenever I knew he was visiting, waiting for him to come in. I'd hear his car revving on the street and watch as he'd drive away. I'd sigh, but sometimes I don't think it was just out of relief.

***

We pull into his driveway.

"The fuck we doing here?"

He shoots me a look. "No one's home. You wanna?"

"Like what—old times' sake?"

He shrugs, flashes a coquette's smile. "Or some-thing."

I smoke a joint in his bathroom. He's sitting on his bed naked, as gorgeous as I remember. Dick sticks up like a fattening pool cue. Huge smile, sun through the window shines on his blond hair like backlighting. Looks like a halo. Even nature's a liar.

I sit down next to him, clench, eyes closed, prepare for a fist or strangle or full-body smothering. Instead his fingers run lightly across my face. I open my eyes, sur-prised at the touch, like feathers.

His hands reach for my shirt and remove it. Lips trav-el across my torso. My pants are off, and now his mouth and tongue wrap around my penis. This is the first time he's done this. Slight jab of teeth. My hands grab his hair, like I'm trying to pull each strand off individually. When we fuck he does it slow and careful, a contrast. I luxuriate in each minute movement—so this is what it feels like when a dick eases into me instead: It's like my ass is a mouth gobbling it up inch by inch. A flush of heat begins in my belly and works its way through my bloodstream, entering my head—this is a fever, or no—I wonder if every inch of skin is glowing, if I could cause an early summer by simply opening my mouth. I want to ask him but I don't want to ruin it. His grunts mix in with my moans, and any word would disrupt the harmony.

He holds me when we're done, my back to his. He whispers into my ear, "What's with all the bruises?"

I tell him about Thomas.

He lives in Pomona. Really phat house, like humungous. Rich shit all over the place, like crystal and china and vases and crap. And all this fuckin' gay-ass furniture, like who would wanna put their butt on something that nice?

I was outside the mall skating, feeling lucky 'cos I found a half-empty pack of Pall Malls, which are totally gross, but beggars can't be choosers. It's hot and I'm sweating so I ride over to the movie theatre exit. I think maybe someone'll throw away an extra-large soda. People always buy the huge one and then drink like ten percent of it.

"Hey mister," I yelled to this total square, walking by himself to his car. "Can I have your soda?"

He sort of started up, body flinched, then turned to look at me. He started to grin, real lecherous.

"You could tell he wanted something," Evan says.

"Well obviously."

"Why would you want someone else's soda?" Thomas had asked.

"'Cos it's hot and I'm thirsty and I got no money," I responded, rolling over to him, passing him, turning around, and skating back.

"He went and bought me a lemonade from Hot Dog on a Stick."

I can tell Evan's listening intently.

"He promised five hundred bucks if he could do whatever he wanted to me. I don't know how you can turn that down."

"You are so beautiful, it disgusts me," Thomas had said as I sat down on the couch, his voice dipping at the contradiction. The hair on my neck stood at attention. "Do you realize how beautiful you are?"

I shrugged, grunted, "I dunno."

"How old are you?"

"Sixteen."

"What I thought. You look exactly like a sixteen-year-old boy should."

"What's that supposed to mean?"

He edged closer, a look in his eye indicating roadside caution signs. "If you know what's good for you, you won't ever grow, or gain weight, or change, or anything."

"I don't see how I can help that."

He shook his head, laughing. "What a mouth. I should destroy you so you won't change. You can remain perfect; they'll remember you fondly. You won't devolve." He wrapped his hand around my neck, and as I attempted struggle, arms weakly pushing against his brawn, he whispered, "Five hundred and I can do anything. Remember."

I almost lost vision as I was losing breath. He let go and I fell forward; his swatting hand helped me collapse to the ground. Trying to get up is almost like a push-up, which I could never do in P.E. A leather-booted foot rushed to my face. I became upended and laid on my back. I couldn't breathe. It was the blood in my nose.

"So he rams this syringe into my neck, and the next thing I know I'm all floaty. My head's light, I can't feel anything, but it's weird 'cos I'm laughing, like it's this total bliss godhead thing. He brings out this knife—"

"Jesus Christ."

He traced lines all over, and if I forgot about the blade it tickled. My eyes wandered and focused on his hand. His muscles began to clench. Harder now, he dug in with the blade. My body looked like a delta, red blood rivulets. He picked up a metal bat and began swinging. The trickling blood became a flood over plain white. My body inched across the hardwood floor, left streaks— paths I'd traveled to get there.

My mouth dropped open, let out a desiccated moan. Didn't sound like much: just a toddler's belch, the air wheezing out of a balloon, the squeak of a door whose hinges are starting to rust.

His dick was hard and dripping. He stuck it in, pulled out, cum filling the gashes on my back.

"Why the hell would you put up with that shit?" Evan asks.

I wanna say, Fucking ironic question.

He gave me a bath, tea, dinner, the cash, bandaged me up, drove me home. He kissed me on the cheek, like he loved me. It didn't seem contrary.

We met two weeks later. I'd semi-healed by then.

Evan proceeds to hold me tighter. "That guy's a pervert."

"No, really?"

"You should stop seeing him."

"I need the money. Maybe I should rob him next time."

Evan is silent. "You know where he keeps his shit?"

"Yeah."

I can feel him thinking. "When are you supposed to see this guy next?"

"I dunno. I could call him, make an appointment."

"Do that. And I'll show up. We'll beat the shit out of him, take everything."

"What if he calls the cops?"

"He's been molesting a 16-year-old kid. He won't do shit."

I turn to face him. We're inches apart. His breath smells like my genitalia. I smile. "I could bring you along, say he can do the same shit to you, and then before it happens we can gang up on him."

He frowns at this. "I dunno if I wanna be in there while—"

"Don't be a pussy, dude. He'll get excited. You're exactly his type."

Evan's brain wrestles with this; it shows on his face. "If you think that'll work."

"It will."

We smile. It's like a handshake. I call Thomas that night, call Evan immediately after.

"Hello?"

"It's a go. Tomorrow night."

***

Evan and I pull into Thomas's driveway. The house looks like any other in Pomona.

"You sure this guy's loaded?" Evan asks. He's been nervous all day. We even fucked to relieve the tension. Didn't work. He's acting all spun.

"I've made 2,000 bucks off him. Yeah, he is."

"Shit. I'm just—what if we botch it?"

"We won't botch it. C'mon."

Thomas is there to greet us at the door. "Well, young sir. It's good to see you again. Been quite some time. And I see you have a friend." He extends his hand to Evan, who looks at it warily before taking it. "Nice to meet you." To me he says loudly, "I see what you mean," like he's Liberace or somebody, and gives a theatrical, faggy wink. "Come in, come in." His arm sweeps toward the open door and he bends slightly at the waist.

He's set up some spread. Food and beer and pot and blow all over the place. "Make yourself comfortable." Evan's eyes bug out at me, and I can only shrug. We reach for anything we can stuff into orifices.

They shoot the shit, it's weird. I know what's gonna happen, he knows what's gonna happen, and yet everything is jovial and cool. It's unsettling. I feel I'm not cut out for this even though I know I have to be, need to be.

Evan is a mess, totally drunk and faded now. I don't know how he thinks he'll be able to do anything. Doesn't help too that Thomas has handed us acid. What the fuck are you doing, Evan, you idiot?

"Oh shit," he says, and looks at me. I don't know if he's aware or if it's just the drugs talking. Then there's this weird-as-fuck noise that blasts through the living room.

"Holy fuck, what is that?" My heart feels like it's trying to escape through my mouth.

Thomas comes in. "Shouldn't you know?" He berates me, sashaying through the room, trying to locate something. "This is that CD you brought over last time."

My ears calm down. It's that Sigur Rós song. "Svefng-englar." It's nonsense but majestic. We both look at Evan, who's totally tripping.

"Holy shit, this is rad, what the fuck is this—whales?"

No, it's a human's voice rendered otherworldly and sort of transcendent. So I guess there was no reason to not take the acid. The sound of the bow scrapes across the guitar strings. Even when you're sober it's like your guts are being ripped out and your mouth opens to vomit intestines, and you're crying from the force, but somehow it's cathartic.

So.

Evan must be losing it. He is. He's standing up, arms wrapped around his stomach, tears brim in his eyes like hastily poured diner coffee. They bug out and he spins his head around fast, body following it, like he's trying to watch something streak across the walls. He looks like a tornado. I think maybe he can blow us out of this, blow the roof off, blow me into the safety of sky. I feel myself flying towards the stars, glancing behind at a mess that's diminishing exponentially till it's a mere speck, disappears. I'm part of a constellation. Orion's belt.

He stops spinning, rocking back and forth from foot to foot, dizzy. His eyes meet mine, his mouth agape. He looks about ready to melt. Or combust.

"I think I'm fucked up, Justin." A whisper.

I nod.

"I think I've fucked up, Justin!" A scream.

We both nod. Thomas takes a Louisville Slugger to Evan's head. The sound of skull cracking on wood, a thump, and then the skull cracks on wood again. Evan, just like that, is out. Blood courses from his once-beautiful face.

There's a crack in the bat. It used to be an impressive object. Everything's ruined now. This almost makes me cry.

Thomas turns to me, breathless, eyes wide like it's Christmas morning. "I think I've got it under control," plunging a syringe into Evan's neck. "Unless you wanna watch. Wanna watch?"

I shake my head. "No, I wanna do it too."

He hands me a knife. We start to dig in. I wanna orgasm after the first plunge. His body is softer than I'd expect.

***

After the third time with Thomas, it felt like a marriage or death sentence; either option involves something approaching honesty. I guess that's what happens when you've seen someone's insides, or you've shown them yours.

Thomas watches me. I'm at the table, wrapped up in a blanket.

"When I was a kid, younger than you," he starts musing to the air, "there was a guy named Randy. He lived two streets away. I'd be playing in the front yard or something, and he'd fly by on his goddamn bike. And then he turned sixteen, got a car. I thought he was so cool. Tall and athletic, blond, huge smile. I wished I could be like him. No one else compared."

"Uh-huh," I grunt. People can get really boring—it's hard to listen to contemplation. Doesn't help that he put on Fleetwood Mac. They kind of rule.

"I'd be walking around the neighborhood, or at the park, and he'd pull up right next to me, call me names, throw shit at me. It was always so embarrassing."

"Sucks, man."

"He beat the shit out of me once. At the park. Raped me. It was almost exactly what I wanted. Then he beat me some more, laughed, then just left." Thomas is raging, his eyes look ready to pour out of his head.

He turns to me. His stare hurts more than anything he's done to me. He holds my face in his hand, says, "It's uncanny."

That "Say That You Love Me" song is like really loud right now. It's so queer but sorta rocking. Weird lyrics actually: "I guess I'm not as strong as I used to be/ And if you use me again it'll be the end of me." Bizarre sentiment for a dumb love song.

He pushes my face away, gets up, walks over to the

couch and throws my clothes over to me.

"I better take you home. I don't think you can handle another session right now."

***

"You know that guy I told you about?" I told Thomas on the phone.

"The one who molested you?"

"Yeah. He's in town. I mentioned you."

"What did you say?"

"Not much, actually. But I think he'd be perfect for you."

"But you're the 16-year-old. I can only use 16-year-olds."

"But he's like me, only more so. He's right up your street. This needs to happen for you."

"I dunno. It doesn't seem exactly right."

I nodded to myself, felt ready to curse. Then I heard Thomas sigh on the other end.

"But I guess this is your revenge, not mine."

Suddenly something inside sparked, and I felt ready to shoot into the air and explode in shimmering color. "It can be both."

"Yeah, I guess I can see that."

"So that's a yes?"

A pause. "Tomorrow."

A yes.

SEAHORSE

by Michael Graves

"God will get you," I whisper.

She's strung in tubes and quilts and corn-colored flesh. Still, my mother is void. She's blank. She's cool.

I had promised Woody one hour each week, and right now, ten minutes remain. So I'm snapping through *Woman's World*.

Hurry. Hurry. Hurry.

Judy, the jumbo nurse, squeaks in. She'd gone to grade school with Mom and is always speckled with fourteen karat gold.

"How's our girl doin'?" she asks.

"The same."

"I know it's hard, Kiddo."

"Sure," I say, turning past a Pampers ad.

"She's in my prayers. Every mornin'. Every night."

I think to myself, "SHUT THE FUCK UP, LADY! You don't know anything!"

"Your Ma really does look terrible."

"Well . . . she's dying."

"No one deserves cancer. Especially in the girly parts."

Know this: I've watched the illness gorge its way through my mother since last June, and as time drips on, I gleam.

"Jesus Christ," Judy soon snivels. "This is just . . . so unfair."

"Um . . . I have to go to the mall."

***

"Mornin'," chimes Clara, the saggy clerk.

Dappled in diamonds of sweat, I veer through checkout lane three. I plunk everything on the speeding black belt. I unload razors and deodorant. I set out a rabbit's foot and five discounted cans of red spray paint.

"Good deals," she says, tapping in each price.

"Yeah."

"Mall's crazy today, huh? Everybody's out gettin' ready for the holidays."

I sigh, but don't speak.

"Havin' a clearance sale next week, ya know."

Then my eyes catch the Speed Stick glowing green. Her register says $2.10.

"Everything's supposed to be a buck," I protest. "This is Buys For A Buck."

"Honey, it's just a name. Can't ya read all these tags?"

"That's retarded," I snip.

"I don't make the rules," the woman says. "Ya want me to ring it or not?"

"No way."

Spurned, oozing with glares, she begins to pack away my goods. "I know why you're always in here buyin' so much air freshener n' spray paint," she tells me. "I ain't dumb. I know ya gonna go shove it up your face. That's what my son used to do."

"I'm eighteen. It's legal."

***

She wheels across the food court, K-Mart bags knotted to her stroller.

Gina's my best friend and one of the only girls I can bear. During tenth grade, I fed her Freeze Pops while she forced out a nine pound baby. We brewed formula, got high, stopped all the screeching.

"Did ya get it?" she asks.

"Yeah."

"No hairspray, right?"

Instantly, I sink to see the beautiful boy. "Hi sweetness."

Jessie just stares at traffic. He gurgles while thick slobber twinkles on his chin.

"Guess what I got," I say, pulling the purple foot free. "Look Jess . . . fancy. It'll give you good luck."

"What's that thing?" Gina asks.

"A surprise. A rabbit's foot. For him to play with."

"George . . . um . . . ya can't give a baby that kind of crap. They eat everything. He'll, like . . . choke."

In a huff, I cock my red head. "I was just being nice. What the fuck?"

"Don't have a hissy, George. Don't be mad. Let's watch all the cute boys."

We both sigh and sit on a graffitied bench. Behind us, the Johnny Appleseed Memorial fountain spurts sadly. It once flowed, gushed, but now it barely trickles.

I start plowing through my pockets, snatching out coins. And I hurl them. Pregnant with hope, I see the pennies drown in dead green waters.

"Make a wish," Gina says.

Know this: God still seeks redemption. He has offered Woody and my perfect new nose. He has offered Mom's death too. So I'm certain the Lord will grant me a pack of sons. Someday.

"Think it'll come true?" she asks.

"Yes."

Gina smirks, knowingly, "Ya could always adopt one. Ya could probably buy one . . . somewhere."

"Maybe I'll just get pregnant."

She gags twice. "If ya do, you'll be the richest, most famous boy in the world."

"I'd love to be huge with a big big belly."

"George . . . ya don't got an egg. Ya need an egg."

"Girl parts are gross," I giggle.

"Ya need a uterus too."

"So. Miracles happen all the time. You never know. What about the Virgin Mary?"

My best friend begins gawking at boys from Burger King. With a purse and a pout, Gina slides on sunglasses.

I tell her, "Come is more important anyway. Maybe all I need is lots and lots and lots of come."

"Betcha already got a name picked out."

"Ace," I say, busting with glee.

"Sounds like a porno name."

"Shut up. It's cool."

But Jessie begins to howl as if he's just had five boost-er shots. He yanks on his shabby mane. Whipping. Wailing.

"What's your problem?" Gina snaps.

Our eyes plunge. The rabbit's foot lies, now drenched, on the glassy floor inside the fountain.

***

I'm at the sink, scrubbing skid marks from his Jockeys. I sing with Rufus, and I'm happy since it feels like summer and I've already guzzled two wine coolers.

Soggy coupons color the counters beside me. Each is cut, not torn. Some perfect squares say, "35 Cents off Pledge," "55 Cents off Luvs," "Buy 1 Six Pack of RC, Get Another FREE!"

Soon, Woody stomps through. About him: He once hid a heap of Honchos beneath the pull-out. After my blustering fits, he now knows not to waste.

"Hey Gorgeous," he grins.

"No *Telegram* today," I say. "Paper boy didn't deliver."

"Did ya call the office?"

"I don't want to get him in trouble. I just want the fucking flyers."

My boyfriend kicks off his beat boots. Squares of hardened mud crumble to the floor.

I'm thinking, "I already vacuumed!"

Then a bitchy smile splits across my face. "Can I get a kiss?"

Woody says nothing, just wiggles from his work

clothes. He blows two wads of snot into the trash can.

"Yuck!" I exclaim.

He's beaming, potbellied. "I'll kiss you now."

"Gross!"

"Come here!"

Woody tortures me with a quick round of tickles. He finally nips my chin. There's bristle and stink and peppermint schnapps.

"Um . . . what'd ya do today?" he asks.

"Nothing. Went to the mall. Saw my mom."

"How's she doin'?"

"Still can't breathe so well. Smells weird too."

His whole face softens.

I tell him, "She doesn't even know I'm there. She's practically dead."

"Really?"

"Yes. So why do I have to see her?"

"Think of it like this . . . if ya go, you'll get sent to heaven."

"I'm already going there."

Woody fake-punches my cheek. "Gotta do what's right."

I begin groping for my can of Kodiak. Fingering out a fat minty bulge, I stuff it in. I suck and I suck.

"Ya still think about all that stuff?" he asks.

"No," I lie. "I don't even remember."

But I do remember month-long earaches. I do remember the olive loaf at Christmas. I do remember all those piercing cavities.

"What we eatin'?" Woody asks.

"Shepherd's Pie."

***

Yes. Yes. Yes.

Red paint flecks dot my slick upper lip. I'm spritzing the washcloth and cupping my face. With a heave, I pull in sweet, clean vapors.

Gina's laughter falls like flurries. More about her: In May, she'd disappeared. Gina met a boy at Bob's and spent the weekend slurping down Buds in his basement. So I took Jessie. When she returned, Gina was grounded for the first time since tenth grade.

"We shouldn't huff no more," she says. "But I love being wasted."

Wet coughs fire from my chest. Sputtering, I belly flop on the twin bed beside her.

She stammers. "Remember . . . a few years back? On that New Years? We drank a bunch of nips . . . and we danced on the highway?"

"I remember."

Gina jolts with a cackle. "George . . . I'm gonna kill myself. I swear. I swear to God," she jokes.

I'm teasing too. "Shut the fuck up."

"My life is like . . . it's . . . the worst."

"Be quiet."

"I live with my mother. Still," she giggles. "I don't got a job. I don't got a boyfriend. I don't got anything."

"You have Jessie."

In a haze, she snorts and wipes away streams of chemical goop. Her smile is suddenly gone.

"I'd do anything to be like you," I say. "To have a little baby."

"Kids can be a pain in the ass, ya know."

I fold my legs. "Just isn't fair."

"I know I'm an asshole. When I think about it . . . if I didn't have a baby . . . I could have so much more." Now she's babbling louder. "And anyways, I ruined the kid's life. It's my fault he's messed up."

My damp eyes begin to flicker because I always forget Jessie's retarded.

***

"Do it!"

"Oh . . . yeah."

Woody fucks me fast on the sofa. He drives in, disappears and then, three seconds later, slides free. Completely.

"I'm gonna shoot," Woody snarls.

"Don't pull out."

"You're so fuckin' hot."

Right now I'm inside a sweet slow dream. Beyond the squeals and hoots, my brain just fizzles. I think of bibs. Booties too.

"I'm gonna come," he gasps.

"Fuck my pussy!"

Woody's drawing on the Lysol-drenched rag. "Ya want my baby batter?"

***

Couples strut across the dance floor ceiling. They boogie, clapping, clapping.

As an old Soul Train jitters, I'm flipped over in the midst of one long headstand. Quick pangs of ache shoot through my head like comets.

"George? George?"

Woody grumbles in. "What the fuck are you doing?"

"They're dripping out. It won't work if they get away."

With a squint, he utters something and bumps back through the darkness.

***

"The eczema's comin' back, darlin'," she says.

Sashes of sun pierce my eyes, but still, between white hot bands, I see nurses smoking out back.

Boring. Boring. Boring.

Judy dumps Jergens in her palm and mashes both hands together. "The cream's so cold," she tells me. "Ya gotta warm it up first."

"Oh."

She begins to rub my mother's hefty arm. Stroking. Squeezing.

Of course, I'm filled with sass. "Why are you doing that, anyway?" I hiss.

"Her skin's dry. Wouldn't you want someone to put cream on you . . . if you couldn't do it yourself?"

I say nothing. I think, "Gross."

"George? Would you do me a favor?" she asks.

"Well . . . I have to leave soon."

"It'll only take a sec," Judy says. Now she's coating her own elbows with leftover lotion. "Why don't ya give your Mom a hug? It's her birthday."

"She doesn't know what's going on," I snicker.

"She knows you're here. She knows. She was blinking and coughing last Tuesday."

Swiftly, the room begins to shrink. Faded flowery walls inch closer, and Judy's just a breath away. I'm being squished.

"Come on, Kiddo. Please?"

I tell her, "We hate each other, you know."

"That's not true!"

"Yes it is. You don't know me. And you don't know her."

The woman quakes. "I know she's a sweetheart. I know she's an angel. She's sick, George."

"So are you," I say. "So am I."

"You're supposed to take care of your Mother."

"She never took care of me!" I shout.

"Settle down. Ssssshhh!"

At that moment, I breach.

"SHE USED TO LIVE IN THE ATTIC, JUDY! AND SHE'D CRY AND CRY AND CRY! All the time! Every day! And she'd never come down!" I sneer, "She'd even pee in Tupperware bowls. And she'd shit in plastic bags. Because she was too scared to leave. She couldn't. Not even on Christmas! Not even on my fucking birthday!"

"Stop cussing! Quiet down!"

"No!"

Judy's wild-eyed, her head bobbing. "You're tellin' stories. And you are on drugs again."

***

I see my cupcakes tan. I sit before the window, looking in, drawing up whipped cream clouds.

And soon, the room flip-flops. In a frothy swirl of giggles, my hump rises, round as a kickball.

It's him.

Ace.

He's growing. He's swimming. He's squirming.

"Hi little man," I gush, so giddy.

***

"Thank God tomorrow's fuckin' Friday," Woody says. Hunched over, he builds a chop suey mountain.

I tip back my wine cooler and tell him, "I'm getting a job at the mall."

"Why?"

"We're going to need the money," I say.

"No we won't. I make enough landscapin'."

"Well . . . kids are expensive. Gotta buy diapers, toys, formula."

Woody stabs the saucy mound. Eagerly, hungrily, he scoops forkfuls into his mouth.

"Have to get clothes and booster shots and . . ."

"Georgie, please." His big blistered hands slice

through the air between us. "Just give it a rest."

For an instant, we stop and our strain soars, floating like a spritz of Old Spice. I can hear the pipes thud. I can hear the Crowleys cursing next door.

So I begin to yap, "We are gonna have a baby!"

"Georgie, you're so messed up. When you gonna realize we aint gettin' a kid? No way!"

I slap the tabletop, and the swift, solid crack stings my hand. "I can do whatever I want."

He glares at me and glares at me.

"I want a baby!"

"Enough!" he thunders.

My fury ping-pongs, bounding throat to womb.

"You can't have no fucking kid, Georgie!"

"SHUT UP!"

"There ain't no miracles."

"SHUT UP!"

"You don't got no cunt!"

***

As curtains of punch-colored clouds tumble above, I sit on Doyle Field's fifty yard line.

Of course, I'm raving. "I should get a job! I should leave Woody! I should move away! I should call Henry What's-His-Name!"

Then my torso tilts to life.

Ace.

I rub his full crowded bump, tracing hearts and swirls.

"Hey . . ." Woody starts to whisper from the sidelines. "I cleaned up the dishes. And the juice," he says.

My belly rumbles around.

"Chop suey was good," he grins. "Think I'll take some for lunch tomorrow."

"Do whatever you want."

Woody drags himself close and drops beside me. "Georgie . . . don't be mad."

"Too late!"

"Why ya wanna a little kid anyway?"

I flick at the wind. "I don't want to be a movie star. I don't want to be a singer. I don't want to be a model. I want to be a parent. That's all I want."

"Yeah?"

"We could raise our kids to feel happy. We could raise our kids to feel safe. It would be fun, Woody."

"It would be fun. But . . . you and me . . . we're only boys. Only eighteen. We can't have no kid. Whose gonna give us a baby?"

"I'll just . . . have it myself," I say.

Blinking at a blank scoreboard, he chuckles. "Right."

"Fuck you." I sneer, snipping reeds of trimmed grass, but in seconds, I toss them like New Year's confetti. "I believe in all these things," I whisper.

He hooks his arms around my waist. "You're my baby. And I'm yours."

With a sputter, I begin to leak into Woody's lap. "Don't you believe too?"

***

Last Christmas, Woody gave the paperboy a card with ten dollars sealed inside, but still, he doesn't deliver and, soon, I just might phone that office.

Right now I'm waiting by the storm window. A Tot Finder decal coils off its glass. The sticker is pale, bleached, baked on by years of sunlight. So I begin strapping tape to each corner curl.

"George? Hey!"

Gina thunders through, spangled in a new crown of sun-splashed hair.

I ask, "Did you see the *Telegram* in the yard?"

"No."

"Fucking A!"

"George! I'm sick," she gripes. "My belly hurts."

"Why?"

"Got my period."

"Eeeewww," I flinch.

Drooping, Jessie lops, lame on his mother's hip, yet in four seconds, the little boy begins to yowl.

"Stop it," Gina crows. She rests him on the floor.

"Hi honey bun," I sing-song in falsetto.

Jessie just licks the linoleum. The baby rolls around, burbling, burbling.

"I put his favorite toys in the purple bag," Gina tells me. "If he gets bitchy, just give him one. He'll cut the crap."

"OK."

"So . . . do I look alright?" She orbits and fluffs her frozen bangs.

"Yeah."

"I can't believe Kurt asked me out. Weird."

"You'll have fun. He's cute too."

"I wish I could fuck him, George. Maybe he won't care. We could put some towels down or something."

Immediately, my forehead slinks toward the ceiling.

"Does that sound skanky?"

***

"Gross!" I shout. "You're smelly."

Jessie's diaper is crammed with loose gobs of shit. His stink strangles.

"Yuck!"

Skating across floorboards, I nab two toys off the table then, glide back.

"Which one you want, Cutie?"

Blankly, he digs at a sore shaped like Florida. The baby's scratching and pulling, dragging and clawing.

"Sweetheart, no."

He grates again.

"You're going to bleed," I say, prying his fingers free.

Colored in spite, the child sniffles.

"Don't cry. Don't cry."

I snag a spit-soaked bunny from the floor and push him toward Jessie's grip. "Look," I offer. "See."

He coos. He slips into silence.

"I'm a pretty good mommy, huh Jess?"

Slowly, bit by bit, the baby lifts his head. As he gawks at me, a real grin curves across his face.

"Soon, I'm gonna have my own baby. A boy. Somebody just like you. He'll be really smart, really fun."

Jessie laughs out loud.

***

See, I often try to forget the unsigned permission slips and each lost weekend. I try to erase my mind, but all the mess remains.

Still, I press her palm to my stomach.

"Feel that?" I ask. "I think there's a baby in there. I think he's really there."

Her cold husky fingers twitch.

"I can feel him sometimes. Mostly in the morning. When I vacuum. When *The Price Is Right* comes on."

"GEORGIE!"

My mother's head springs up from her pillow. "GEORGIE? GEORGIE?" she croaks. "I heard ya. Can ya hear me?"

Skimming backwards, I knock the tiny table and a mound of old *McCall's* slip to the floor.

"George?"

I'm draped in wonder. Awestruck too. "What?"

"They're comin'. They are."

"Oh yeah?"

She starts to yank on the hissing hoses. "They still find me. Every night they come."

"Still?"

"I'm scared," my mother moans. "I'm so scared."

"Well . . . soon you'll be dead."

"I'm sorry." Now she's throttling with tears. "Please don't tell your baby . . . about the things I did."

It feels as if God just pressed pause because life outside is still and the nurses don't gibber.

"Tell him I was . . . magnificent."

My mother cannot see the sheen of dribble that gleams on my cheeks.

"I will."

***

I'm tipsy with hope. I swat through a CVS sack and scoop out my E.P.T. The box screams, "A woman's FIRST choice . . . CLEAR RESULTS . . . 99% accurate."

In a glance, I see Clara, from Buys for a Buck. She waddles across the food court, hugging her handbag.

Quickly, I stash away the test.

"Hi," she says, and slumps down beside me.

I say nothing.

Clara tells me, "Waitin' for my son, Derrick. Should be comin' soon. Said he'd be by round noon . . . but it's almost one."

I think: "Who cares?"

"Maybe he forgot," Clara says. "Derrick always forgets."

I say to myself, "Be quiet! Shut up!"

Then she thumps my arm three times. "Hey . . . uh . . . we got air freshener on sale, ya know."

"Oh."

"Some cinnamon spice stuff. If ya like that kind."

Looking at her, I hold my stare and, slowly, I tell Clara, "I quit doing that."

But a wet sour mist peppers our skin. There's slopping and cackles and clucks all around.

We watch school skippers hunt through the fountain. They're sweeping up handfuls of coins.

"Lord," Clara croaks. Bumbling, shaking, the old woman starts to rise. "Those are people's wishes! My wishes!" she hollers. "Now they aint gonna come true!"

***

It's dark enough to switch on every lamp in every corner, but I don't because it feels snug. It feels secret.

I'm rocking in our beaten La-Z-Boy, whisking batter for another carrot cake. I dunk my middle finger and lap the candied goo.

Beside me, Jessie's still snoozing. Soundlessly.

Then a burst of thuds ruptures at the front door. Cradling my bowl, I shoot up and unclick each deadbolt.

It's Marion, Gina's mother.

Fuck. Fuck. Fuck.

"Where is she?" the woman asks, rubbing her weary golden eyelids.

"On a date."

"I don't think so."

I blend three slow circles. "Maybe they went to the mall or something. Maybe they went to get subs."

Marion says, "She took her make-up and some blouses."

"So?"

"She took her birth control too."

"Oh."

With a smirk and a sigh, she unbuttons her coat. "Did Gina tell ya about the pills her uncle bought over the computer? The Zoloft?"

"No."

"Well they ain't working so well."

"She'll come back. Soon."

"If ya see her, tell her I'm tired of her shit, George. I gotta work, ya know. I have my own life."

"I'll keep Jessie. No big deal."

* * *

"Is this gonna be OK?" he asks.

While my brows knot like a birthday bow, I dress us all in afghans. "What do you mean?"

"I dunno. We've never slept like this . . ."

The baby is wrapped, bundled, nested between us. He's already dreaming.

"Should I put a shirt on?"

I shake my head, thinking, "Woody . . . you're so cute."

"What if I crush the kid?" he says.

"You won't. Jessie'd probably scream anyway."

He rolls closer and jabs a finger in his bellybutton. Woody sniffs it. "Ya think Gina's comin' back this time?"

"Yeah. By tomorrow. By Friday, definitely."

With a slow, drowsy wink, he folds the blanket over Jessie's shoulder.

"This is how it could be," I tell him.

***

See: I promised God that if Ace comes home, all will be settled. I'll be nice too. I'll donate to the State Troopers Fund and sweep the Crowley's crumbled walkway. I'll even attend mom's funeral.

Right now my legs are unlocked. As I drench the test wand, specks of urine hit my hand.

Please. Please. Please.

The cordless starts to sing its droning song. There's a crackle, then Woody's voice on the answering machine. "We'll call ya back," he echoes.

"Hey . . ."

Gina.

"Don't pick up. If you do, I'll hang up."

Scowling, I bunch a wad of Charmin. I dab at my penis.

"I'm in Atlantic City," she says, beginning to sob. "I know that you hate my fucking guts. But I can't do it! I gotta be . . . alone. Away from my mother. Away from Leominster. And Jessie. And you too."

I swish back down the hall and perch atop the whirling washing machine.

"I'm sorry. I know I'm a piece of shit."

Behind her blubbers, a girl shouts, "TWO HUNDRED BUCKS! TWO HUNDRED BUCKS!"

Gina's groaning, "I called my Ma. I told her that you'd take Jess. I know you'll do that for me George. You'll be a great Ma. Better than me. Better than . . ."

The box beeps.

Suddenly, Woody booms in. His face is crinkled and he clutches today's *Telegram*.

"Hey Gorgeous," he smiles.

My eyes dip down to the stick still in my hand. I can see a bright red plus sign, blurry and perfect.

I'LL PAY YOU BACK WHEN I GET HOME

by Michael Wolfe

When I turned thirteen, everyone in my life vanished. My father to his secretary's bedroom, my mother to church groups and the casino, my sister to a college dormitory, my brother to delivering pizzas and drugs, and my first boyfriend, red-headed Tommy Bachman, into thin air. This was 1984; we were both thirteen and paperboys for *The Des Moines Register*. I worked because Dad insisted I learn the value of a dollar and needed the exercise, but Tommy worked so he could spend more time with me.

I quit my route the day after Tommy disappeared. When I walked to the corner of Ashworth and 29th to pick up my bundle, I saw a picture of Tommy smiling at me, folded on the lips, under the streetlamp. Beneath him sat 526 more. I lay down and held the stack and cried.

Every day after we finished our routes and before we went home, Tommy and I played Frogger across Ashworth Road, the busiest road in West Des Moines. He lived on the corner of Ashworth and 27th, and near the sidewalk stood a brick wall as tall as us. Sometimes we sat on the top of the wall and pointed to passing cars we planned to buy after saving enough from our routes. Other times we threw our extra papers at cars from behind the wall until one of us made contact, and when we saw brake lights, we'd book as fast as we could up to Tommy's bedroom and lock the door and peek out the upstairs window before we fell asleep under his brown afghan. For Frogger, one of us hid behind the wall while the other stood across the street giving hand signals. You couldn't see the oncoming traffic from behind the wall, so you'd have to look across the street for a go signal, and once that arm shot in the air you had to run for your life before you got smashed. I was fatter and slower than Tommy, so I always cheated and stood where I could see the traffic, and I'd make up excuses for not running when he told me to. He'd never talked about running away without me.

Later that morning, Dad took me to his office in the skywalk because I was too sad to go to school. I hated everyone. I sat in a leather chair and watched him call clients and memorized combinations of numbers and letters on the electronic stock ticker circling the office walls. Dad wasn't a real stock trader on the floors of Chicago or New York; he wore a suit and tie, and all he did was talk on the phone.

For lunch Dad took his secretary and me to Lombardo's, an Italian joint in the skywalk with a decent view of downtown Des Moines. Sheila laughed at everything Dad said, even if it wasn't funny, even when he asked the waiter for more water, even when he told her my best friend was missing. When she laughed, her tongue peeked out of her teeth, and every time she spoke she'd just trail off to nowhere or stop mid-sentence.

Maybe I wasn't listening very well that day, but I don't remember much of what she said.

"Levinston Brothers is a solid firm," Dad said, stabbing his fork into a cherry tomato. "Good place to work."

"Bo-ring," I said.

"Think you'd like to take over my books when I retire?"

"Can I have a quarter?"

"Gotta earn it," Dad said. "Now what can we have you do for a quarter?" He looked at Sheila, who giggled into her straw.

"It's just a quarter," I said. "I'll pay you back when I get home." I'd only pushed around my lasagna—it looked like spin-art. I touched the tip of my tongue to my nose.

Sheila blew iced tea through her straw, over the table and the arm of my father's blue shirt. Dad winged his elbow out to look at the stain. "Tell you what," he said, standing up. I could see the worried lines of his fingertip. He breathed hard and excused himself.

Sheila handed me a quarter. I walked to a pay phone in the skywalk. Mrs. Bachman answered. I covered my free ear to hear her because the lunch crowd had filled the skywalk.

"Tommy honey, where are you?"

I held my hand over the receiver and listened to her cry. Before I hung up I lowered my voice to tell her he was OK.

Back at the table, Dad and Sheila fed each other pieces of tiramisu. I counted aloud the time it took each of them to fork a bite into the other's mouth, and then I started over when one of them swallowed. One one-thousand, two one-thousand, three one-thousand, four . . .

***

Dad shut his office door and talked seriously.

He picked up the brass bull-and-bear bookends I had given him for Christmas. "Do you know what depreciation means?"

I shook my head.

He steadied the bookends in mid-air. "Let's say this bear here is your mom, and that I'm the bull."

"So."

"So there's a bull market, and there's a bear market." He lifted the bull's horns into his chest, just below the knot of his red tie. "You either bull up," and then he put the bear in front of his fly. "Or you bear down. And right now your old man's in a bear market." He faced the animals towards each other on the top of his desk. "And it's not your fault. I don't want you to think this has anything to do with you. The economy's a shit storm right now, and Sheila's a good woman. You got that?"

The door opened behind me. Sheila leaned her head in.

"Everyone's got a worth, son," Dad said. "Never forget that."

"What are you?" I asked Sheila.

She looked at me dumbly.

"Are you a bull or a bear?"

She half-laughed and threw her hands in the air. Then she finger gunned me.

"A big, fat bear."

***

Even weirder and fatter without Tommy by my side, I returned to school two days later; the teacher, Mrs. Patton, left his desk next to mine, and it still had all of his stuff in it—crumpled notes and rubber cement balls I'd given him—and during recess when all of my classmates were outside, I opened Tommy's supply box, took out his compass, and then carved the number of days he'd been missing, 3, into his desktop.

My school counselor, skinny, faggy Mr. Tillman, asked me how bad I missed Tommy (on a scale of one to ten) and how mad I was at my parents (one to five). I didn't care about my parents because they didn't care about me. The last night Dad slept in our house I'd overheard them discussing my brother and sister, and they had said, naturally, Cliff would go with my father and Kristin with my mother. They didn't say my name. All Mom said was that he'd better not bring Sheila to Mom's church. And that she'd hired the best. She wanted half.

Mr. Tillman made me show him my sketchbook.

"Are these pictures of you with your father?" he said, and flipped the pages quickly.

"No." I took a mint from his candy dish.

"It's very natural to miss your father in a time like this." He compared two of the pictures, and then looked up at me. He was so gross to me; I didn't want his fingers on my sketchbook. "Can you tell me why all of the men are different looking, but the boy is always the same?"

"They're perpetrators."

I'd been taping *Unsolved Mysteries* and watching the tapes late at night while Mom gambled or prayed or gambled and prayed, and Cliff delivered drugs. I'd sit on the couch with a bag of Cheetos and Tommy's art supplies, which I'd lifted from school. I had taped over Kristin's choir concerts and Cliff's little league videos because it was one less thing my parents would fight over. I paused the tapes at each composite sketch and then copied the faces into my sketchbook. After I ate enough Cheetos, I drew Tommy next to each perp, smearing my fingers on the page to give Tommy his red hair. Sometimes, I rubbed my fingers through my own hair.

Mr. Tillman told me I wasn't worthless, but he couldn't tell me how much, exactly, I was worth. I wanted a dollar amount.

After school I walked down shady Vine Street, stopping at Casey's General Store for a newspaper and a pop and a bag of Cheetos, and took them to the metal picnic bench under the storm shelter at Fairmeadows Park, where high school kids smoked cigarettes and kicked hackie sacks in the parking lot and two soccer teams practiced on the field and a few mothers pushed their children on swings. I recognized some of the high schoolers as my brother's drug buddies, but I had never really talked to them. They were a combination of hippies and punks, and one of them, who everyone called Lizard, opened his car doors and turned up a scratchy Grateful Dead bootleg. He was skinny, bald and tattooed, and drove a wood-paneled station wagon covered in bumper stickers. And he talked the loudest. I wondered how much a tattoo of Tommy's name would cost.

I unfolded the newspaper on the picnic table. It had only taken three days for Tommy's disappearance to be buried in the Metro section on page 5. The Des Moines Police Department was offering a fifty thousand-dollar reward for any information on Tommy's whereabouts. Anytime Mrs. Bachman was quoted, it made me hopeful that he hadn't simply vanished, but that he'd just taken a long bike ride without me. I checked the stocks page for Berkshire Hathaway's price, which Dad had said was one of a kind. One Tommy was worth more than twenty shares of Berkshire Hathaway.

I unzipped my backpack and pulled out my sketch-book and a pencil. I began adding the prices of everything I owned. I spent an hour calculating the prices of my baseball card collection, clothes, and mostly third-place trophies. Then I unzipped the front pouch of my backpack and counted the wad of savings from delivering papers. By the time everyone but Lizard and I had left the park, I was worth four hundred and thirty-three dollars and sixty-two cents, and I was bawling.

Lizard danced alone in the parking lot, his arms flailing in the air. He was older than my brother and most of his friends but still hung out with high school kids, even though he had his own apartment and dropped out three years prior. Cliff told me he was looped—like really fucked up from all the acid. So fucked up that none of the chicks would date him. He was a runaway and a mooch. I threw my sketchbook in my backpack, washed my eyes with water from the drinking fountain, and walked towards the parking lot.

Up close, Lizard's skin was seriously scaly, peeling like silver crayon from a lottery ticket. Most of his arms were covered in black and blue tattoos, and the only one I could make out was the black goat on his neck with horns wrapping around his head. I just stood and stared at him.

He stopped dancing and leaned into the car to turn off the stereo.

"You're Cliff's little brother," he said, tapping a cigarette out of a soft pack.

"Not anymore," I said. "I'm nobody." I was about to cry again. I felt it in my teeth.

"Listen, I'm friggin' bust this week. Can you help me out? I'll get you back next time I see Cliff." He struck a match against a front tooth to light a cigarette. He held the pack out in front of me. "Smoke?"

Something about him made me want to try it, but I said no. I didn't want him to see me cough.

"All I have is $433.62."

Lizard held in a drag and looked up at the sky. He closed his eyes, exhaled, and froze for a moment, like he was praying. I wanted to run. I thought maybe he'd seen me counting my money on the picnic bench and that he was considering whether or not to punch me and steal my backpack. I took a few steps away from him.

"Cliff thinks you're a pretty cool little bro. Not ratting and shit. Suppose you've been bummed since your friend disappeared."

"I don't want to talk about it."

"This old hippie once told me it was all porn rings. That these creeps take kids and change their identities and brainwash them and then make them do all this whack sex shit. I met him on tour with the Dead. Flipped my freak out. He gets me all stoned in his van and locks the doors and pulls the shades shut, and then starts telling me this shit and I'm like, Bro, this isn't cool. And I swear to God it's the never-ending joint. I can't get out of the van. Hands down, worst high of my life. Missed the first half of the show because of that fucker. And they opened with Jack Straw and went into this sick version of Loser. I've got the boot but it's never the same as being there. You know? And get this —they close with Tom Dooley. Can you fucking believe it? Sure, I see the second set, but that cat had already sincerely messed me up. I thought he'd murder me."

I sat on the curb with my face pressed in my backpack.

"May the four winds blow him safely home," Lizard said, and put his bony hand on my fat back.

I jerked away from his hand and stood up. "Don't touch me," I yelled, and ran away from him as fast as I could. He called after me apologetically, but I kept running.

My shoes were untied and I had a cramp, so I stopped on Fairlawn Drive. It was getting dark. I was hungry and a few miles from Mom's. I didn't feel like being anywhere. I wanted to call Kristin at college and have her come pick me up. Or go to Tommy's house and sit in his bedroom while Mrs. Bachman made a grilled cheese sandwich. Or jump in the back of a garbage truck and be dumped at the landfill. I zigzagged the sidewalk and looked at the tops of elm trees and chimneys, talking to God like he was Tommy.

"Hey, kid," I heard from the street. Lizard had rolled down his window and turned his lights off. In the dark, he looked like Skeletor. "Hop in."

"Leave me alone," I said, and walked faster. Tommy, seriously, if you're there, please tell me where to go, Tomato Head. Give me a sign.

He coasted along beside me with his elbow propped out.

"Bro, I'm sorry," he said. "Just popped into my head. Happens when you're high. You think everything's worth saying out loud."

"Take back what you said about the sex stuff."

"I take back what I said about your friend. He's not in

a porn ring. He's probably at his aunt's house. Now get in the car before someone sees us. I got a quarter ounce of dope in the glove, and I can't be getting busted."

"You swear you're not a molester?"

"Jesus Christ," he said. "I don't have time for this shit."

I had nowhere else to go. I opened the passenger door and set my backpack on his floor. Lizard grabbed a handful of cassettes from the seat and threw them into the back. Then he pushed in the car lighter and flipped a cigarette between his lips.

"Don't take me home," I said, and we sped off.

***

Lizard's apartment walls were lined with tie-dyed tapestries and stolen street signs. I sat on a broken recliner while Lizard ordered Chinese delivery. Above my head hung a black tapestry of Jim Morrison holding a snake out in front of his bare chest with the words "I am the Lizard King, I can do anything" scrawled across the snake's body. On the coffee table sat empty baggies, hemp twine and a box full of beads. It looked exactly like my brother's room. Even though Cliff dealt drugs, Lizard's apartment was the first place I ever saw or smelled marijuana.

Lizard pulled a silver tray out from under the couch and dumped a bag of pot in the center. Some pieces he studied up close, and other pieces he ripped apart. I didn't exist when the pot was in front of him. I asked how many tattoos he had.

"Fifty-three," he said, and looked down at his right arm, as if checking to make sure none had fallen off. "I get fifty-four next time I have the dough."

"Crazy," I said. "My dad just turned fifty-four." Then Sheila's age crossed my mind. I guessed her at thirty, a twenty-four year difference. "How old are you?"

"Nineteen," he said, sprinkling pot into a thin patch of paper.

"I'm a prime number, too," I said. "Unlucky thirteen."

"Trippy," Lizard said, bowing his head.

"It's the worst year ever," I said. "Did Cliff tell you my parents are getting divorced?"

"Gotta keep on keepin' on."

Lizard slowly licked the full length of the paper before twisting it shut. Then he stuck the entire joint in his mouth, pulled it out, and held a lighter a few inches beneath it. He set the tray on the coffee table and lit the joint. It sizzled and popped. Lizard looked at the lit end and then blew smoke at it. A piece of ash fell.

He put on The Smiths' "Heaven Knows I'm Miserable Now." He fell onto the futon opposite the broken recliner and closed his eyes. He inhaled every ten seconds, exhaled every fourteen, and ashed on the carpet. My stomach grumbled.

"Why do they call you Lizard?"

His eyes shot open. He leaned toward me and took a drag. "Because—" his voice was high and pained, like Mom's. He raised a finger and held the smoke in. It slowly crept out of his mouth and nostrils until he coughed the rest onto my face. He looked to the tapestry above my head and extended both arms, palms up. "I can do anything."

Lizard setup the Atari for me to play before he showered. I tried snooping in his bedroom, but the door was locked, so I sat on the carpet up close to the TV and played *Frogger*. The blood splats on Lizard's television looked real. Tommy was a pro. When we played Frogger, he ran across Ashworth Road every single time.

After dinner, I asked Lizard to pay me back for the Chinese food that came while he was in the shower, but he claimed to be flat broke.

"How do you make money?"

"Creative endeavors," Lizard said. "And I have kind friends."

"This one's on me then," I said. "But you have to get the next one."

"How much money you say you had?" Now Lizard was playing with his pot again but not smoking it. He rearranged it endlessly on the tray like a Rubik's Cube.

"$418.10, after dinner."

"Think you can help a brother out? I'm dry as bones and rent's due soon."

"It's all I've got," I said. "You have to earn your own. Or else do something for me."

I thought I could live like this for a while without anyone noticing. I knew nobody would miss me tonight, anyway. They were all in their own little worlds. I could drop out of school and devote my time to finding

Tommy, and I could start delivering papers again or help Lizard with his creative endeavors, and once Tommy returned I'd have enough saved for my own place. Living with Lizard was the first step to disappearing completely.

"Let me stay here," I said.

"I don't know. Call your brother to see if it's cool."

"No way we're calling Cliff. He can't know I'm here. Ever."

"I will if you don't float me," he reached for the phone.

"Don't." I grabbed his skinny arm.

"Let go of my fucking arm," Lizard said. "And give me the fucking money."

***

Two days later I had thirty-six dollars left, and Lizard's cupboards were empty. Tommy was five days gone. Lizard smoked pot all day and only left the house in the afternoons to go to the park. I mostly played Atari and listened to his bootlegs or watched TV. Lizard had sworn to me that he'd hold up his end of the deal and not say a word if he ran into Cliff. I wondered how long it would take for anyone to notice I was missing. Two days in, I didn't feel any different, but I had this weird feeling that Lizard had chained and gagged Tommy in his closet. I finally busted into Lizard's bedroom when he went the park.

A waterbed with a green comforter divided Lizard's room. The rest sat spotless. There wasn't a poster on the wall, not even a wrapper on the floor. It's like we lived in two different worlds, me on the smoke-stained futon, and Lizard in the hotel down the hall. When I opened his closet doors, I found his shirts hung neatly and arranged by color. On the floor lay a stack of *Playgirls* and new tennis shoes. I checked the drawers under his bed and the tops of his closet shelves, looking for ransom notes or a clump of Tommy's hair or his seashell necklace or anything, but I found nothing.

I was sitting on his bedroom floor, looking through the stack of magazines, when I heard the front door. I quickly zipped my pants and shut the closet door. The bedroom door was off hinge, so I propped it shut as best I could.

Lizard sat on the futon, reading the newspaper. No one had reported me missing yet. It had only taken Tommy's parents an hour after their dog, Bondo, showed up on the porch without Tommy. They went looking for him, but only found his wagon on the corner of Orchard Drive, half-full of undelivered papers. I'm the last person they checked with before filing a missing person report. I knew we needed money. At least I did. If I didn't get out of Lizard's fast, he'd molest me or turn me in. I estimated everything in his apartment: bootlegs, futon, waterbed, hemp necklaces, *Playgirls*, television, Atari, posters . . . Lizard's belongings equaled roughly two thousand dollars. I couldn't imagine what all of his tattoos had cost him, or pot. All I had was thirty-six dollars, a backpack, the dirty clothes on my back, and a nearly full sketchbook.

"I think I know of a creative endeavor," I said to Lizard. "That could make us a lot of money."

He grinned and said, "Lizzie likes creative endeavors." Then he scrunched his face. "What's your magic?"

"You hold me hostage and ask my parents for money."

"Ransom?"

"That's what I meant," I said. "I'm pretty sure my dad has more than my mom. And he invests other people's money in stocks. He's a moneyman."

"Oh hell yeah," Lizard said. His eyeballs were red. "I like this. I like where this is going." He pulled his pot out from beneath the couch and a pipe from his pocket. He pushed pot into the pipe with his thumb and then took a hit. "But we gotta think about this. How do we get the money without me getting busted?"

"You ever met my dad?" I said.

"Parents are Lizard's worst nightmare."

"Then you're the middle man," I said. "And the kidnapper. You call and demand the money, and tell him to meet a guy with tattoos at the park. Tell him if he involves the police or the press that you'll kill me, and have him killed. Once he hands over the money, the guy with tattoos, you, will deliver the money and bring me back to the park. Only you and I will know there was never a real middle man."

It made perfect sense to me, but it took two more bowls for Lizard.

"Are we talking thousands or millions?" Lizard said.

"I don't know."

"Check this—they held Patty Hearst at six million. Probably before your time—she's the daughter of this super loaded newspaper guru. I met a guy who got her high one time. Crazy cat from Maine. So after Papa Hearst forks over the money, the kidnappers still don't give her back. Doesn't reappear until she's caught robbing a bank with her kidnappers. You believe that shit? Six million and they try and rob a bank. Greedy fuckers, bro. Gotta know when to split."

I'd heard the story referred to on *Unsolved Mysteries* and knew the money hadn't gone to the kidnappers, but to the homeless in San Francisco. I thought about a life of crime with Lizard and how we'd get busted because of his tattoos or my slowness. I couldn't even finish the mile in gym class without walking. It made me happy knowing I'd never have to do that again.

"So William Hearst is loaded, and his daughter's at six million, I figure we ask what, say a hundred grand for you?"

I knew Dad didn't have that much. "We have to start lower. Dad always says to buy low and sell high."

"How low?"

"Start at fifty dollars."

"Are you out of your mind?" Lizard said. "Then it's over. Fifty bucks is barely a quarter ounce. We have to start out of reach and then bargain."

"You owe me," I said. "Trust me. He won't bite."

**

The first day, Dad had thought Lizard was asking him to buy a single share of stock in my name, and he hung up the phone before Lizard could make a threat. So after that I wrote down exactly what he should say to Dad, and we spent the entire afternoon acting out different scenarios. I was always my dad.

I hadn't left the apartment in over three days. I wore Lizard's patched jeans and old concert T-shirts and I'd solved *Frogger* and *Donkey Kong* and *Q*bert*. I'd learned not to answer Lizard's phone because it only rang early in the morning or near dinnertime, and came from a bill collector.

We reached five grand by the end of the week, and Lizard was so stoked he splurged all weekend, charging new video games and toiletries and concert tickets. But when we hit ten grand the following Tuesday, I started thinking maybe Dad really loved me. Ten days divided by ten thousand dollars meant I was worth a thousand dollars a day. After watching *Unsolved Mysteries*, I asked Lizard where his parents lived.

"Haven't talked to them in years," he said, threading a purple bead onto a hemp necklace. "Could be dead for all I know."

"Do you miss them?"

"The grass ain't greener, the wine ain't sweeter, either side of the hill."

I'd heard so much of his music over the past few days that I knew half of what Lizard said came from song lyrics. Sometimes he made sense, but this time I wasn't sure what he meant—who the grass was and who the wine was and which side of the hill we were on.

"Ten thousand is a lot," I said. "Let's take the money and run."

But that night I couldn't sleep. I worried about Lizard going to jail and how much trouble I'd be in once my parents or Cliff put it together. I knew we could probably get more out of Dad, maybe even triple up, but I also knew stocks only rose so high so fast, and that sooner or later the novelty wore off—markets froze or crashed unexpectedly, companies got bought out, merged, went bankrupt—and so did people.

I turned on a light and flipped slowly through my sketchbook. On a fresh page, I drew a picture of myself with the number of days I'd been gone, 10, above my head, and then I drew Tommy beside me underneath a 13. I cracked Lizard's door just enough for the hall light to shine at the foot of his bed, where I sat cross-legged with the sketchbook on my lap. I drew the comforter's folds over his body from the feet up with a green Conté crayon. His pillow covered his head, but I could still hear him snoring little hisses. When I got to the base of his neck, I finished with the green oil and began with the black. I drew a goat head and then the thick horns. I couldn't decide whether or not to sign my name or write a Grateful Dead lyric or thank you or what he should say to my dad the next day.

I pulled my pants down, sat on the edge of Lizard's

bed, and stroked myself with patchouli oil from the nightstand. Occasionally I whispered his name to make sure he wasn't awake. When I finished, I tore out the page and wiped myself, crumpled it, and shoved it inside the pillowcase. I grabbed Lizard's "Steal Your Face" T-shirt from the closet, gathered my things, and left.

I walked the streets until morning, dodging head-lights and keeping an eye out for bundles of papers. I wanted to grab one before they were delivered to check the market outlook and to see what my parents and sib-lings were saying about me. I was officially missing, but Tommy had been gone so long he even disappeared from the paper. I knew soon that Tommy would have a picture of me to cut out of the paper and keep in his pocket. Once I made it to Des Moines' city limits, I'd stick my thumb out on the side of the freeway, or find a small café to sit down in and rest. I could draw portraits for *Unsolved Mysteries* or write Letters to the Editor with secret messages for Tommy or get a job as a busboy. Or maybe I'd end up just like Lizard. I didn't need any money. I could do anything.

WOLF TEETH

from *Eat When You Feel Sad*
by Zachary German

Robert comes home from work. Tom, Paul, and Chrissy are in Robert's room. Robert feels happy. Robert thinks "My room is cool. I am cool." The song "A Minor Place" by Bonnie "Prince" Billy is on, and Robert feels happy. Robert enters his room. Robert says "Hiya."

Tom says "We saw *Snakes on a Plane*."

Paul says "Hi." Chrissy is on Robert's computer. She puts on the song "Heartbeats" by The Knife.

Tom says "You should play the live version of this; it's so much better."

Chrissy says "He doesn't have it."

Tom says "You suck, dude."

Robert says "Yeah I know." Robert feels happy. Robert takes off his hoodie. Robert hangs up his hoodie. Robert looks at Paul. Robert says "How was *Snakes on a Plane*?" He says the words in an unusual way, in an attempt towards comedic effect. Tom laughs.

Paul says "It wasn't very good." Paul looks at his shoes. Robert looks at Paul's shoes. Robert feels sad.

Tom says "Yeah it wasn't like anything really. I thought it would be a lot funnier."

Robert says "Huh."

Tom says "It had a few funny jokes in the first five minutes that just repeated for the rest of the movie."

Robert says "That's cool." Robert lies on his bed. Robert sits up and takes off his shoes. Chrissy puts on the song "Shoulder Lean" by Young Dro featuring T.I. Robert feels okay. He says "What's everybody doing tonight?" There is a pause.

Paul says "There's a party." Paul and Chrissy tell Robert where the party is, and that it's for Paul's friend's friend, who just got into med school. Tom says he's too tired and that he has to work tomorrow.

Robert says "Yeah I'd go. Yeah alright." Tom, Paul, and Chrissy leave Robert's room. Robert changes his clothes. It is summer. Robert goes outside, and Paul and Chrissy are smoking on the stoop. They finish smoking and say goodbye to Tom. They walk to Chrissy's car and drive to the party. In the car, they listen to a commercial hip hop radio station. On the way to the party, they stop at the beer store. Chrissy gets a forty of Milwaukee's Best for Paul and a six pack of Pabst tall boys for Robert. Chrissy gets a forty of Milwaukee's Best for herself. At the party, Robert says "This party is funny." The music is loud.

Paul says "Yeah." Robert, Paul and Chrissy drink their beer. Robert drinks two Pabst tall boys, and Paul and Chrissy each drink three quarters of their forties. Paul talks to his friend whose friend's party it is for a little while, and then they leave. They ask someone where the nearest Chinese food store is and then they go there. Robert buys a Phillies Blunt and a homemade iced tea and a vegetable lo mein. Paul gets a vegetable lo mein. Chrissy gets a homemade lemonade. While they wait for their food, Robert smokes his Phillies Blunt. Paul and Chrissy smoke cigarettes. Robert's lungs hurt. They eat their food on the steps of a house near the Chinese food store. They walk to Chrissy's car. They drive to Robert's house. Tom is watching the David Bowie video collection with Brian. Robert, Paul, and Chrissy tell Tom and Brian about the party. The David Bowie videos start to be of songs that Robert doesn't know.

Robert says "I don't know these songs. These songs are bad."

Tom says "Yeah. Do you want to watch something else?"

Robert says "I don't care." Brian suggests that they watch a foreign film, *Little Otek*. Brian and Tom look for it for a little while, and then they find it. Tom puts it in the VCR. Robert has a hard time reading the subtitles. He is sitting near Paul. He is drinking a Pabst tall boy. He asks Paul if he wants one.

Paul says "Yeah, I think I would like that." Robert feels uncomfortable. Robert thinks "Everything is wrong. I hate myself." Robert feels drunk. Robert thinks "I have been thinking this for a while, about what I am going to do. This is right." Robert thinks "I hate myself." Robert watches the movie for a little while. Robert finds some paper. Robert writes I LIKE YOU AND WANT TO BE BOYFRIENDS. Robert shows it to Paul. Paul says "Yeah, I . . . feel that way too." Robert feels happy and nervous. Robert feels nervous. They watch the movie for a little while. Robert feels tired.

Robert says to Paul "Do you want to go upstairs?" Paul
nods. Robert and Paul go upstairs. Robert brings his two
remaining Pabst tall boys. Robert thinks "They think I am
gay. No they don't. I don't know what I am doing." They
go to Robert's bedroom. Robert feels happy that he has
more beer. Robert and Paul lie down. Robert says "Do you
want to split a beer?" Paul nods. Robert feels nervous.
Robert opens a Pabst tall boy. Robert sits up in his bed.
Paul sits up in Robert's bed. They drink beer. They talk
about feminine beauty and sex and what they are doing.
Robert thinks that maybe Paul doesn't know what Robert
meant earlier. Robert doesn't remember exactly what he
wrote or where that paper went. Robert feels nervous that
someone downstairs might look at it. Robert says "I've
never done anything like this before." Robert finishes the
can of beer. He puts it on the floor next to his bed.

Paul says "Like what?" Robert cries a little. He doesn't
think Paul notices.

Robert says "Like with a boy. It's a lot different."

Paul says "We're not really doing anything."

Robert says "Can I kiss you?"

Paul says "OK." Robert kisses Paul on the lips. Robert
feels Paul's facial hair and wants to throw up. Robert
knows that he can't do this. Robert puts his hands on his
chest and turns away from Paul. Robert thinks "I am
horrible." They talk for a little while longer about how
this is different. Robert tells Paul that he is really beautiful. Paul tries to get Robert to talk about Alice. Robert
thinks that he is going to throw up. Paul says "I think I

am going to go."

Robert says "I'm sorry." Robert thinks "I am glad that he is leaving." Paul gets out of Robert's bed and turns the overhead light on. Paul puts on his shoes. Robert is surprised that both of their shoes are on the floor near Robert's bed. Robert says "Are you OK to walk?"

Paul says "Yeah it's only like five blocks." Robert gets out of his bed. He doesn't hear the TV.

Robert says "Do you want me to walk you home?" Robert feels stupid.

Paul says "No Robert." Robert feels sad.

Robert says "OK." They go downstairs together. Robert says "Do you want anything? A glass of water?"

Paul says "No." There is a pause, and Robert thinks that Paul is going to say something else. Paul opens the door and steps outside. Robert steps outside. It is warm. Robert thinks "I am somehow surprised that I can see so clearly."

Paul says "I'm going home now."

Robert says "I'm sorry."

Paul says "It's OK." Paul walks away. Robert goes inside and closes the door. Robert goes upstairs. He goes on YouTube and puts on the live version of "Heartbeats" by The Knife. Robert gets in his bed. Robert calls Alice. She doesn't pick up. Robert gets out of bed. Robert puts his head next to the wall until the song is over. Robert puts on album *I Can Hear the Heart Beating as One* by Yo La Tengo, starting with track seven, "Stockholm Syndrome." Robert gets in his bed and lies awake for two hours. Robert goes to sleep.

BRAD PITT

by Wilfred Brandt

I mean, who in their right mind would say they'd had a bad time on a tropical island resort with Brad Pitt? No one. That's why Marnie didn't. Even though she did.

Her first day back at work, she hunkered down amid the stronghold of her cubicle, harboring a brave face and blotchy sunburn. Coworkers at Mitchell Stanley Auto Insurance arrived throughout the morning, bundled up against the Midwest winter. The females all approached with wide eyes and raised eyebrows, eager to know, "How WAS it?!?" And though the men's grunted inquiries were more discreet, they were equally curious and jealous.

How could she tell them that really, it kinda sucked? The food was bad, the resort was tacky, and the company . . . ?

To begin with, Marnie wasn't really a "beach person." Travel agencies pimp out scenes of seaside serenity, and music video rock stars parade 'round St. Bart's like it's "the bomb." But too much sun had Marnie in hives, seawater made her hair explode in afro frizz, and sand irritated her sensitive Irish skin.

Marnie just couldn't understand the appeal. What few trips the Minnesota native had made to proper "beaches" had been awkward affairs, one hand securing floppy hat to head, another clutching tightly to a sarong or some such covering, spitting bugs and sand and wiping sunscreen-soaked sweat from her eyes, waiting for the "relaxation" to end.

But when Brad Pitt invites you away for a week's holiday at a tropical resort, you can't really tell him you're not a "beach person." Well, I mean, you CAN, but you wouldn't. So she didn't.

Their destination was a fancy resort in a third world country, the kind of place where dark-skinned natives hawk their centuries-old culture like so many puka shell necklaces and low-cost pearls. Marnie felt strange being one of dozens of slightly overweight, distinctly un-tan white people fawned over by an array of natives twice her size with half her income. Marnie hadn't traveled much—she hadn't even had a passport before this trip. She wondered if she'd feel more comfortable in Spain, or Greece, or the south of France, where she would at least (sort of) blend in with the locals. But who's going to look this kind of gift horse in the mouth? So she smiled politely as she sat in a golf cart beside the man voted *People Magazine*'s "Sexiest Man Alive" in both 1995 and 2000, her hands folded in her lap, whizzing past smiling dark faces raking lawns and cracking coconuts and doing other menial tasks for minimum wage.

The resort was pleasant, a smattering of sleek modernist white townhouses with tasteful timber flourishes and top-notch steel appliances interspersed amongst ridiculously well-manicured lawns, sub-divided by pebble walkways lined with speakers generating light, innocuous electronic dance music (unlike what Marnie was used to but certainly not offensive). Marnie thought it would be a nice place to retire to, or die in, or both. Brad seemed to be in his element. Then again, Marnie was sure this was the sort thing Brad lived with every day. Being a different class of citizen was his very existence—like it or not—and his daily routine must consistent of countless interactions based in servitude and jaw-dropping, nervy "I'm not worthy!" reactions.

And then there was Brad. How could Marnie put it? Oh, he was sweet and gorgeous and charming and lovely. He was certainly more romantic than Oscar, Marnie's partner of eight years and recent ex. Oscar's idea of a romantic getaway was a night spent side-by-side with two buckets of coins in front of the slots on the Riverboat Gambler. But while Oscar was full of outlandish dreams that never materialized and erratic behavior that never disappeared, Brad turned out to be . . . a little dull. Making conversation over breakfast, lunch, and dinner was grueling. Marnie's life as a filing clerk was far less interesting than that of a movie star, she'd assumed. But she was having second thoughts after hearing all the minutiae of Brad's life: his dogs, his new haircut, his battles with irritable bowel syndrome, his elaborate moisturizing and exfoliating routine, the difficulty of finding a good reliable personal trainer, his thoughts on the housing market in Southern California, his current spate of home renovations, his disinterest in contemporary fiction, why he hates coriander, etc, etc.

To clarify, it was not that Brad ever asked about her life. He didn't. But not because he didn't care—he did. He doted all over her like crazy. It just never seemed to cross his mind. Marnie quickly realized Brad was painfully self-involved and a bit of a bimbo. He didn't know who Winston Churchill was, for Christ's sake. This discovery led to one of Marnie's many stints of awkward silence, as her attempts at leading towards meaningful conversation (and away from idle chatter) ran into yet another dead-end, and she gave up all hope, albeit temporarily.

One night Brad tried to use "wilty" as a word in Scrabble. Which, when you think about it objectively, isn't really a big deal. Hey, if Brad Pitt wants to whisk you away to a five-star tropical island resort and make up words to win at Scrabble—so be it! But when you've had to endure five days of elaborate moisturizing routines and avocado-scrub-mask evenings, and when you've had to scour your brain to come up with engaging small talk that doesn't gallop over a fourth grade level, and when you've been made to endure countless "dutch oven" practical jokes (which Marnie had never ever heard of before meeting this charming Brad), it ain't easy. Brad kept insisting, "used in a sentence—my salad is a little 'wilty,'" but Marnie wasn't buying it. Besides, Marnie's upstanding nature and independence were two of the things that attracted Brad to her, so she certainly wasn't going to let a "wilty" slide, especially with the triple letter score on "W" ensuring that the game would be Brad's.

All this blonde bimbo business presents itself as hot property in a Calvin Klein print ad or a Levi's commercial (remember that one with Brad, sans shirt, donning cowboy hat, backlit and sweaty? HOT). But when you're face-to-face with the embodiment of all things young, dumb, and full of cum, it's distinctly less appealing. Sex is cerebral, and while Marnie didn't wanna do it with Brad's brain, she at least needed to know he had one in order to get excited.

Brad had B.O. And bad breath. And was a lousy kisser. And he loved to talk dirty and get Marnie all worked up. Or, more accurately, he liked to whisper breathy, vaguely erotic yet oddly clunky (or creepy) phrases into her ear, in an attempt to get her hot. Which had the opposite effect. Example: "Oh yeah baby, you want me to do a little cunn-a-lingus for you?" How do you answer that with a straight face? Actually, how do you answer BRAD PITT asking you that with a straight face? Or "Your panties are all wet, like a cheerleader —are you a lost little cheerleader? Where's your ride home?" Eww! Marnie was thinking, "oh for-the-love-of-God, Brad Pitt would you just shut the up and screw me," and then she couldn't believe she was thinking that, and then she couldn't believe this was happening, and then she couldn't believe what a bad time she was having while this was happening.

Waking up next to that man whose *Fight Club*-pumped abs made millions of women sigh simultaneously is one thing. Waking up next to that same man to have him lay in bed for 45 minutes trying to decide what to have for breakfast is another. Marnie rattled off a cattle call of every conceivable breakfast option. Brad just wasn't sure what he was "in the mood" for. Did he want potatoes or brioche? A hot meal or the "consequential" (he meant "continental") breakfast? As her suggestions were met by more and more ludicrous refusals ("Mmm, I think of pancakes as more of a winter food," "Bacon makes me phlegm," "I'm off 'white foods' at the moment"), Marnie sincerely wanted to club the gorgeous Greek god on the head with an oversized chocolate brown throw pillow. Everyone has moments of indecision, but most people's moments clock-in well under five minutes. Brad's had exceeded thirty. Marnie wasn't hungry when she woke up that morning, but by the time they left their sleek, modernist townhouse—an hour later after more of Brad's puttering—she was starving. And pissed-off. And over it.

How do you tell anyone your holiday away on a tropical desert island with Brad Pitt was frustrating, uneventful, and boring? You don't. You say that it was great, the time of your life, unbelievable, awe-inspiring. And deep down you wonder if maybe there's something wrong with you. Or maybe there's something wrong with Brad Pitt. Or maybe there's something wrong with all of us. Oh well. At least Marnie got the frequent flier miles. Maybe she'd see if Oscar wanted to go somewhere nice next spring.

SPUNK

from *Hyde Street*
by Stephen Boyer

Wet, rapping Beastie Boys' "Sabotage," dancing alone in the shower, my trance began: the grisly babysitter my parents constantly hired opened the door and asked to take my cock, small pubescent bush, and inexperienced cum loads in her mouth. She was the first person I ever touched sexually.

"All the kids at school have already tried," she said.

Her clothes slipped off her body, then she climbed her fat self into the shower, got on her knees, and gulped my pecker. I began masturbating a year before, so I knew what fucking was. Watching her slurp my dick surprised me past the point of caring to wait for someone physically attractive to be my first. Under my bed, I had a collection of porn that I stole from a video rental store next to my house. I knew what was and wasn't attractive. People in porn seemed much more talented than she was, I thought, as the surprise subsided and my desire intensified, leading me to need an actually attractive dripping wet orifice. She was hungry chewy, whereas porno people appeared to be hungry tycoons out to own, out for everything. I was twelve.

At the end of the night as my mom drove her home, my babysitter told Mom she got a real job. The zit-covered, sex-curious fat girl never showed her face again. I came in her throat, didn't bother to warn her; she must have been seventeen.

When I turned seventeen-and-three-quarters, my mom caught me jacking off. Originally I was supposed to be getting head from my "best friend" Tim, but luckily for him he was running late. Dad would have punched us both in the face if Mom had walked in on us sucking each other dry. Fortunately, I sat on the couch working my dick to a lipstick lesbian gang bang—six girls on vacation in a country cabin eating muff and strapping on dildos all weekend long. Tim and I liked to swap and take turns role-playing as women giving men blow jobs. We had a G-string we'd alternate wearing and a blonde wig. He liked it when I rubbed his back and told him how great I thought he was. I liked to put on lipstick and tell him to pull my hair.

Mom and Dad came home an hour after they left because they forgot the tickets to a corporate "We're going to save the world and eat expensive cheeses on a yacht" event they had scheduled.

"Oh my God! John!"

I ran down the hall toward my room, away from the garage where Birth Dad stood slouched over, cumbersomely getting his large body out of the car. I slammed my door and put on a pair of sweatpants. Mom said she couldn't look at me, so my dad cracked the door and said, "Tomorrow, we'll talk."

The next morning I awoke to him sitting next to my bed, rubbing my back, a cup of water in his hand.

"Your Mom and I have decided, we think it's best for the family for you to get some help. We're going to have you move to a Christian Correctional Facility until you are cured from this disease."

Since Jesus hates masturbators and I am a masturbator and I was days away from turning eighteen, I decided it was my time to venture off into the world and disconnect myself from my parents' idle reality.

"I'd rather go stay at a friend's house," I said. "Could you give me an hour to pack up some stuff?"

"No."

"What? I need my stuff. You can't take my fucking stuff."

"Don't you DARE use that tone of voice with ME."

I rolled my eyes and moved toward the closet, his big blue eyes staring, believing they were powerful enough to conquer me.

"I'm not staying here. You and mom can't peer into my private life like that. It's creepy."

"Creepy?"

"Haven't you read *1984*?"

"What are you talking about, creepy? YOUR MOM WALKED IN ON YOU MASTURBATING."

"I know. So I'm leaving."

"NO. Can't you listen? YOUR MOM AND I TALKED and we decided you are going to go to stay at the clinic and they will help you get better."

"I'm almost 18."

"And that is why you will go to the facility. Do you

realize what would have happened if your sister wasn't spending the night at her friend's house and happened to walk in on you?"

"No. I wouldn't have done it if she was around. I'm not a moron. I should be able to have control over my own life."

"Nahhhhhht."

"Yessssss. I'm eighteen years old, and that means it's completely normal and healthy for me to be masturbating. HAVEN'T YOU TAKEN HEALTH CLASS?"

"DON'T USE THAT TONE OF VOICE." He got right in my face.

I backed into my clothes, cowering into my favorite black pea coat. "FINE."

"What have we talked about all these years? Since you've been an infant, Stephen, I've done my best to instill good Christian values in you."

"I didn't mean for Mom to see me."

"Ultimately it doesn't matter if Mom sees you or I see you or whoever . . . GOD SEES YOU."

"It's not like I got off on you coming home early. Hormones weren't my idea."

"WHAT?"

"I'm not going to that facility. It's final."

He kept arguing with me, neither of us willing to give up our resolution. "Stephen, we love you, but we can't tolerate this kind of behavior."

"Whatever. If you loved me you'd accept me."

"We do accept you, you are my son. We love you."

"Accepting I am your son doesn't mean you've accepted me."

My mom started to cry as my dad's authoritarian dictatorship continued its eradication of masturbators from its territory.

Maybe the tears are proof of love, I thought, then started packing the few things I knew I'd need: jacket, sweatshirt, two pairs of pants, three pairs of socks, underwear, three shirts, a pair of boots, sunglasses, deodorant, toothbrush, toothpaste, facial scrub, facial cream, small towel, laptop with headphones, and a copy of *The Complete Collected Poems of William Carlos Williams, 1906-1938*. As far as non-religious/public-school academic writing was concerned, at the moment of leaving home I had only really read and understood a tiny bit of Allen Ginsberg's *Howl*, but my English teacher instructed that I should begin looking at the intersections of structure, rhythm, and image in Williams' work. I thought I'd give it a try. Before packing my computer into my bag, I put an ad on Craigslist. I looked at several ads before placing mine, Google searching the lingo to make sure I depicted myself correctly.

Subject: Young Versatile Stud Needing Live-In Situation

Body: 6'1" 165lbs, 8.5-inch cock, brown hair, brown eyes, slim/average build, versatile, able to suck you dry/I never run out of cum/18 years old. Far from home, need live-in situation, drug and STI free, take care of me and I'll take care of you.

After placing the ad, I put my computer in my bag and walked out of my parents' front door. I didn't bother to say anything to them, since I never wanted to speak to either monster again.

My first night on my own, I looked up at the smoggy sky and knew death would one day come. Am I denying Carl Solomon and Allen Ginsberg by running from my Rockland—the Christian Correctional Facility? I thought, as I climbed onto the roof of my high school and slept next to a heater vent. As the warm steam lightly wrapped itself around my shivering body, I realized I hated Ginsberg for wanting to return to Solomon in Rockland instead of pulling Solomon away. I always thought Ginsberg understood life is about ejaculation, not suffering. I've had it with living and dying (later in life, I see he is Buddhist).

I didn't let myself cry because my parent's Christian idealism never acknowledged my emotional life during childhood. It taught me to remain quiet, inward, and detached from the outside world, which kept me naïve. As the night progressed, I realized I didn't care for Ginsberg any longer because he was a poet, and poetry is what led me to feel comfortable as a masturbator, which led me to sleep on a rooftop, which led my eyes to open for the first time to the darkness I would have otherwise stayed uninvolved with had I remained in my parents' gloom. Despite my resolution, the cold night entrenched me. The detached, inward, quiet being I had been brought up to be broke and screamed, "Fuck you motherfucker!"

In the morning, I drank coffee at a little café with free Internet. I thought about what to do, where to go, who I could fuck, knowing I was liberated to sell myself to the highest bidder. I got an e-mail from a man in San Francisco. "Older generous gentleman here, and you're exactly what this hungry cock needs," he wrote. Could have been worse, I thought, then I responded with, "Fly me to San Francisco as soon as possible because I need to get out of my life before my dad kills me."

**

"Hey baby," he said as he grabbed my bags.

"Hi. I'm Adam."

"Welcome to San Francisco. The car is over here."

"Cool."

I didn't really know what to say as we walked to the car, so I kept quiet.

"Over here," and he motioned me toward a mid-'90s pickup truck. He threw my bags in the back as I hopped in.

"So, if you don't me asking, what exactly got you in this position?"

"My parents are really religious and they kicked me out," I say.

"Their loss, my gain," and he laughed, thinking he was a clever, funny man.

I laughed too, but my laugh was part of the service I offered.

"What you like to do, baby? You know, you're totally hot."

"I like to read and draw a lot."

"You like to read and draw?" and he again erupted like a hyena. "Baby, I mean, what you like to do in bed? What gets you off?"

"Oh. Um. I really like to get blown. Sometimes I like to blow, depends on the guy though. I just recently got fucked for the first time, it kind of hurt but I kind of really liked it at the same time. It was weird. You?"

"Have you ever been tied up before?"

"No."

"You haven't, baby?"

"I've seen it in a movie."

"Call me master, baby. I'm going to teach you how to take life by the balls and nail it."

We were driving fast down the freeway as I related what was about to become my reality to a porno flick I was vaguely remembering. We crossed through Golden Gate Park on 19th Avenue and ended at Cabrillo and 23rd Avenue. He parked the car, and we walked into a large, white, three-story flat. I climbed the stairs to the third floor and entered my new three-room, two-bathroom chamber. The master bedroom was our shared enclave, room one was his office, which was strictly forbidden for me to enter, and room three was in the center of the apartment and set up as a dungeon. I had never actually witnessed a dungeon before entering his, so he had to explain to me what exactly I was going to be doing as the rest of the world hurried past.

"What do you want to use as a safe word, baby?"

"Fire."

"OK. Baby, now just take as much as Daddy wants to give you."

He stuffed his hands inside my ass. I thought about crying out for God, then I thought, What else or who should I cry for? Police, aliens, the president, a dog, a friend, a banker? Each seemed the better alternative as I moaned, "Fuck, oh God dammit, shit." Looking back, I know I was stuck in a K-Hole. Instead of killing my father, I let him push me out into a repeat, more honest version of the same relationship.

Birth Dad is such a moron; he thought kicking me out would get me on my feet. Instead, I arrived in Mr. Old Cock's arms, and he demanded I hang upside down from his bootstraps. Ecstatic waves of adrenaline poured through my body as all the blood rushed to my head and his paw smacked my ass. Fucker needs to get shot, a part of me thought.

"Fire."

"What baby?"

Goddammit, I love sex so much, and this was my first time actually living as a sex worker, which is the job of lower class kings. HIV was my only fear. Was it a rational fear or a government conspiracy unable to attack me because Mr. Old Cock was a recent retiree from the government? I freaked out, imagining his jizz as he flogged me. He poured a bucket of warm olive oil all over my body then jacked me over and over again.

"Aaaagh!" I yelled. Killer. He was a killer, and he

owned my body. Lips on my cock. Marked by his paw. None of this mattered because I was his slave, and slaves aren't people, slaves are dirt. Only humans matter, and the way I was treated when we fucked wasn't human; it's like I was a minority, except I was not a minority. Technically, he was a minority, but he had money, which meant I was a minority. All the racism bullshit school taught me was crap. Pansexuals are just as good as heterosexuals, but if I were pansexual with a billion dollars, I would've been more human than most humans. "Queer" is just a term used by people too busy talking and not fucking. But when we weren't fucking, he was the perfect gentleman, offering me gifts and telling me to sleep as he worked from the forbidden room. Responsible sex didn't get me off as quickly as sex without condoms, and we never used condoms because he was my master, and master won or else I got whipped.

I quickly realized I'd be whipped regardless. Seriously, I wanted to be human and I wanted to understand the pleasure he got from taking complete power over me, but I couldn't. Tycoon is all I saw, grubby old tycoon who wanted me to watch him eat and rub his hairy, fat belly afterward. I underestimated the fucker, I know, but I've since met more bears who aren't as tycoon as he was, which leads me to see he cannot be considered a stereotype because there is no such thing as a stereotype unless your brain doesn't work. Vice was all I was wanting, and he got it. When I get on my feet, I'll never do this to anyone, is what I thought the first time we played, but the second time I decided I wanted to feel the sense of power and pleasure he was enjoying as he beat my puny insignificance. Xenophobia ruled my master's sex life because he is a sexually vengeful Native American; spanking me and shoving his cock in my throat was karma's way of avenging his forefathers, he said.

"You're insane," I said, but it didn't matter. I had no money, and if I moved out, I'd have to live on the streets because, without proper credit, Americans do not exist. I learned that while I learned I am not a human. Did the forefathers ever consider that some Americans might lose their rights to the demands of machines? Zilch is worse than pain. I learned that from my Grandma when she told me Great-Grandpa spent all the family's inheritance.

***

(712) 433-3369: Hi this is Michel.

(712) 489-2140: Hi. I e-mailed you pics and you responded.

(712) 433-3369: Who are you?

(712) 489-2140: Stephen.

(712) 433-3369: You sent me pics for a shoot?

(712) 489-2140: Yes. My name is Stephen, but I go by Dorian.

(712) 433-3369: Dorian?

(712) 489-2140: Like the Oscar Wilde character, but without the last name. Well actually you can call me Slay.

(712) 433-3369: Dorian Slay?

(712) 489-2140: Yeah. Do you like to read? Have you read . . .

(712) 433-3369: Honey, I live in a world of images.

I held the phone tight to my cheek picturing what he might look like. Who sat naked in front of him? My cock got hard imagining a life without words.

(712) 489-2140: Whoa. Not reading seems so weird.

He laughed. Paused. Then said, "I remember your e-mail, exciting pics. I'll see you Monday at 11 a.m. sharp."

(712) 489-2140: OK.

(712) 433-3369: Don't forget to bring three forms of ID.

(712) 489-2140: Can't wait.

(712) 433-3369: Alright. Chat later.

***

For the interview, I wore black, tight-fitting jeans, a grey T-shirt, and black boots. I didn't really know what to expect or what to wear because fashion never seems to be prevalent in porno. When I first walked in, the receptionist asked me to sit on the couch next to the door. Nervously I sat and took out Charles Burns' *Black Hole* from my bag to read, hoping to show that I wasn't just a raw lamb chop. When Michel walked in, he took me to a back room and asked me to strip, so I did, hungry for his approval.

His lazy eye focused in on nude me. He asked me to twirl in a circle, and I did. Besides his lazy eye, he's a total typical middle age alt-fag: forty something, tubby belly, precise exaggerated speech, balding, one hoop piercing in his ear, white Bauhaus T-shirt with a black leather vest. I stopped as he commanded, fidgeted as he snapped photos.

"Where did you get your tattoo?"

"I got it from a friend about a year ago. In his living room."

"Hip."

"Yeah? I drew it."

"Nice."

"The guy who did it was totally awesome. He was my friends' friend. He learned to tattoo in jail."

"Wild."

"He said he sold machine guns in Nevada and stole cars to get back and forth. Since he sold the guns across state line, his case went to the Feds and he had to go to prison. He learned how to tattoo for crack."

We both laughed.

"What a mad boy."

"He did it for free, too. I was 16, so I couldn't get a tattoo legally. I paid for a bunch of booze, but I think getting tattoos at peoples houses is the only way I could get a tattoo."

"What exactly is it?"

"A Satyr. Like Dionysus's bodyguards. Can I smoke? He's the ancient god of sex and booze. Love that!"

"Me too. That's fine. Can you sit as you smoke and spread your legs for me?"

I lit a cigarette and laid down. One leg up. Chin up.

My dick and balls spread across the floor. "Are the photos turning out OK?"

"The next Jeff Stryker. Can you get hard for me?"

"I'll try."

"I'll help."

I smiled and tugged my cock. Spit in my palm and tugged some more. Getting hard is easy.

"What kind of porn are you really wanting to be doing?"

"I've been having lots of S&M sex lately. And I've been bottoming a lot. I want to get fisted eventually. But I don't know if I'm ready for that."

"Enchanting. Read Foucault."

"Who's that?"

"If you want to get fisted, read Foucault. I'll have the receptionist spell it out for you."

"OK."

"So for starts, we'll do a basic fucking scene, then go from there. No drugs when you come in. Not even weed. I've had peoples eyes get all weird on camera or freak out from the lights, and I don't want to deal with it. No alcohol. Shower before you come in, like right before you leave the house, and I do not want to see any body hair. I like my boys smooth. Except have your pubes kind of there. Like trim up your pits and get rid of most your hair so it looks like you're just starting to transform into a stud. And then from there we'll see where your chemistry leads you. Can you handle all that?"

"OK."

"How big a dick can you handle?"

"The biggest thing I've ever had is a dildo like—" I use my hands to measure out nine-inch length with the diameter of 2.75 inches.

"OK, size queen. I'm thinking I'll get this boy Mark, he has like a twelve-inch dick and he's about 6'2" with a super toned hot body. You handle that?"

I sat up attentively and smiled. I hoped my body and face were capable enough to transform me into a star.

"Maybe you've seen him? Mark Wood? Totally your perfect entrance."

"Sounds hot."

"Good. Next Friday?"

"OK."

"One-thirty?"

"Cool."

"Booked."

Despite my bad tan line, I'm naked on a bed. Mark, a tall blonde dude with muscles bulging out of a white tank top, walks in place for two moments, then enters into the shot.

I say, "Hey, are you here to help me get back at my boyfriend?"

"Oh yeah, baby. What's his name?"

A wide shot is used so we are both seen.

"Bobby." I look directly at the camera, which focuses in on my face, and wave. "Hi, Bobby."

Immediately, Mark starts sucking me and I scream, "Oh fuck," then he starts to stick two fingers in my ass.

I squeeze my head with my hands and moan, "Oh my God, fuck, oh God, oh yeah."

His tongue follows his hands. He screams, "You sexy ass" into my ass as he pulls out for breath. After this we flop, and I start deep-throating his 11.2-inch cock, Lucifer. "Gag on Lucifer," he says, "Yeah, uh huh that's fucking good . . . mmmm." He squeezes my head and continues talking, "Suck it right. I got a reputation."

He pulls it out of my mouth and smacks Lucifer all over my face.

"Bobby is gonna love this," I say, but I'm gagging, vomit building in my esophagus.

"I like it nasty. Uh huh. Say hi to Bobby."

Moments pass and I'm in doggy style taking his huge cock, yelling "Hi, Bobby!" into the camera. I've never taken such an enormous cock in my entire life.

"You like that in you?" The camera focuses on my back and his muscled chest and arms. He smacks my ass as he fucks me. I push my face into the mattress. The camera zooms past my gaping hole and focuses on his throbbing, gigantic cock.

"Why don't you fuck him like a cowboy?" The director, Michel, interjects.

"Have you ever been fucked like a cowboy?" Lucifer's mouth asks.

"No."

"Show him."

"OK. Sit on top of me and face the camera, not me . . . Dig it?"

"Oh yeah."

"Take it all."

"Oh God."

"Oh God. You're so fucking hot."

"Sweat all over me. You're dripping all over me. Your boy pussy is so fucking tight."

"Yeah? You're going to make me come."

"Tight little pussy."

"I'm gonna come from my ass."

"Take that cock."

He grabs my chest and pulls me hard into his dick. He pulls me harder and harder, faster and faster. The room and everything in my brain begins to spin, collides, breaks into rays of light bursting through space.

"Bobby fuck you this good?" He slows down, grabs down on my shoulders, leans back and steadily pumps me with Lucifer.

"I'm not thinking about Bobby right now."

"That's my boy."

"I'm going to come!" I begin to shoot all over the place. I'm barely even hard, but I'm shooting across the room. "I've never shot from ass."

"You are such a slut."

He smacks my ass, then grabs my shoulders so I can't stop fucking him.

"Feel it."

He pushes me off of him and jacks his dick over my face.

"Oh yeah baby, I'm coming. I'm coming for you. I'm

droppin loads . . . oh fuck yeah"

I open my mouth as his thick, gummy cum plops into me, a hungry cannibal savoring his favorite snack.

"Fuck yeah."

"Big smiles boys. Gorgeous. Big smiles," the director says, "OK . . . cut!"

Afterwards, we smile at each other and make our way to the food table. I guzzle a Red Bull and munch on a granola bar. Michel walks up behind me, handing me a towel and promises to use me again. I make my way toward the dressing room.

Lucifer is sitting in there, soft. "What's your real name?"

"Mark."

"Oh yeah. I'm Stephen."

"Oh. OK. Great shoot."

"Thanks. I enjoyed it too."

"You can hop in the shower first."

"Thanks."

I get in and soap myself up. I wonder if he's going to come in and fuck me harder than he already has. I picture fucking without a camera, bareback. I start washing my hair while I watch Mark, facing the mirror, picking at his face with one hand and looking at his cell with the other. He's older, maybe 27.

"Do you want me to keep the water running?"

"Yeah. Do that."

I get out and begin to towel off. Slowly at first, cause I still have hope he will call me in and stuff my mouth with his cock. I finish drying off, dress as slowly as I can while picturing his veiny cock smacking my face. Sucking on it. Him coming all over me. Smacking my ass. Fully dressed, I walk to the front office to collect my paycheck.

"Heard it was a great shoot," the receptionist says, "Here's your check."

"Thanks."

"I'm sure Michel will want to direct you again."

"Cool. Just call me."

I casually make a half-ass attempt at a peace sign, glance at my check, and make for the exit.

Once on the street, I look again at my check and call Tim, who's still stuck in our hometown.

(712) 489-2140: Hey. So I totally did it. I just made my first porno.

(214) 213-2144: Did what?

(712) 489-2140: Shot my first porno. It was . . .

(214) 213-2144: Oh my God. Your mom is gonna

(712) 489-2140: the best sex I think I've had in

(214) 213-2144: kill you when someone finds out about

(712) 489-2140: awhile. He fucked me like a cowboy, I

(214) 213-2144: it.

(712) 489-2140: had to face the camera and sit on his cock and bounce on top of him. I've never fucked anyone like that before, it was amazing.

(214) 213-2144: Did it hurt?

(712) 489-2140: Not really.

(214) 213-2144: Are you going to see him again?

(712) 489-2140: I don't know. He was totally hot.

(214) 213-2144: Cool.

(712) 489-2140: How have you been?

(214) 213-2144: OK. Your parents keep calling me for information about you. Jason was telling me at school that his mom was saying the church is having a weekly prayer group night for you right now.

712-489-2140: Great.

214-213-2144: I know. So retarded, they don't have a fucking clue where you are, so they sit around and pray. HAHAHAHA I want to go tell them what you just did.

712-489-2140: DON'T!

214-213-2144: HAHA I'm not a moron, I don't want to get punched in the face by your dad when I tell him his son is a faggot.

712-489-2140: Thank God. What's happening at home?

214-213-2144: Just lots of school stuff. Nothing fun or exciting. I haven't been seeing anyone.

712-489-2140: You going to come visit?

214-213-2144: I want to try in a few weeks, I just need to get my brakes fixed so I can drive.

712-489-2140: Cool.

214-213-2144: How's the place?

712-489-2140: Fine. It's fine. I've just been reading a lot and watching movies. Got a fake ID too so I've been . . .

214-213-2144: Awesome. Can you get me one?

712-489-2140: Probably.

214-213-2144: How much?

712-489-2140: Mine cost 100.

214-213-2144: Cool. Does it scan?

712-489-2140: Duh, bitch. Can I call you back later I need to go to the bank?

214-213-2144: Love you.

Two and a half months later I got an e-mail from Michel asking for my address. He said he wanted to send me a copy of the DVD. I never got the DVD, nor did I ever hear from him again.

LOVE FOR SALE

by Anthony Paull

I can't find him. The man I've heard about. The man they tell me is out there, the man who I was born to love. I've searched. I promise. Now, it's his turn to find me.

Clever me, I made it simple. Last week, I bought an ad in *The Weekly*. Real estate section. My house, my heart, they're both on the market—what's left of them. The house you can buy for plenty. The heart though, that comes for free. Consider it a package deal: a real bargain for any man capable of loving me and only me. A real deal for any man capable of bringing home a paycheck, and not an STD . . .

The last man, he brought home a rash instead of cash. And the man before that, he rarely came home at all. Said he preferred the company of a nightcap over sleeping next to me.

"Was he stupid?" friends would ask.

"Stupid enough to be an alcoholic," I'd respond.

Me, I don't touch the stuff, though Momma might tell you different. Shed her a minute, and she'll tell you that I must have been drunk the day I devised my plan to land a man. That or I was pumped with painkillers . . .

I assure you, these allegations hold no truth.

You see, my kind of pain can't be killed.

God, I wish it could . . .

If only there were some magic pill, some magic way to ease my sorrow. To ease the emptiness of easing my way into an empty bed each and every night . . .

Maybe I should be grateful. Empty nights are what lent time for me to brainstorm. My plan: simple. Dust off this old house, this old pile of bricks, and lure in a man with good old family values. Respect, fidelity, trust—that's what I'm after. That and a man with a healthy appetite for good old-fashioned love.

The kind of man who will fork you all night and spoon you all morning, that's the man for me.

Momma says I must be damn stupid to believe a man like that still exists. Momma says I need to read more, that men like that have relocated to some other planet. Mars or Venus, she once read. But she can't remember.

She sure can remember how to be mean though, calling me stupid. Just like that mean old real estate man. I heard him whisper the word "stupid" the day he snapped my picture in front of my house for *The Weekly*. I heard him, all right. That mean old man, he kept waving me away, telling me he needed a full shot of the house. But I never budged. No sir. I told him. "I'm on the market too. We have to show the buyer all the amenities."

You see, they're not only purchasing a recently reno-vated two-bedroom pile of sticks. No sir. They're also purchasing a three-dimensional man. A man full of every color of the rainbow:

Green for the naïveté of a man who believes in the "one."

Red for the blood drawn after the last "one" left.
Blue for the "one" long overdue.

***

The first day of my open house is on a bright Sunday afternoon. Here Mr. Real, the real estate man, finds me sunshine yellow, bright-eyed, and glowing with the thought of selling my home and heart to someone worthy. "You have company," he says, knocking two times on the front door.

"Be there in a moment," I call.

Mr. Real, he's really in my face as I open the door some more. His pointy, gray mustache nearly clips me. "Got us some friendly faces," he says with a cigarette smile.

Cracking the door a bit further, I view the friendly bunch: some white, white family with some white, white manners. The parents nod, the children smile, then in unison, a hello. "What a nice surprise," I say. Me and my bullshit smile.

"They told me, 'We want a shiny green lawn for the kids,'" Mr. Real snorts. "And I told them, 'Come with me. I know the shiniest green lawn in town.'"

The kids are eyeballing the lawn. And mom, she's eyeballing dad, as if to say, "Mighty green isn't it?" Me, I'm all eyes on Mr. Real, the really, really mean real estate man. "Can I have a word with you?" I ask him. "Alone."

"Sure," he says, with a plastic snort. Then insta-salesman, he immediately turns to reassure the family. "Go on. Try out that lawn. Feel all that green. Like standing on a million dollars."

The family, they ooh and ahh, taking off their shoes to feel the soft, green grass on their toes. Me, I haven't the heart to tell them what I use for fertilizer.

"Now, how can I help you?" Mr. Real asks me.

"I told you. Find me a man who wants a family, not a man who already has one."

"Now, hold here a minute," he huffs. "I'm here to make a sale."

"Well, I'm here to make a family," I reply. "Go on. Find me a single man."

Slamming the door, I'm back in the hollow of the house. Here I am, alone with that empty feeling. That feeling like a real family should reside here—one that never truly experiences the cold. A home where the fireplace is always lit and the scent of comfort arises from cookies baking in the oven.

From the guest room, Momma hollers that the only thing baking in this house is her big old butt. "Boy, will you turn down that heat?" she asks. Momma's been living here for three months now. She's in remission. She had cancer, cancer of the big old butt hole.

She says I gave it to her. It's my fault. The day I told her I prefer men was a month prior to the day they rerouted her intestines like a highway. Now, she poops off the turnpike.

We don't waste poop here though. No sir, not in this house. The market here, they charge big bucks for animal discharge. Cow manure, Momma manure—it's all the same thing. Green, green dollars to fertilize green, green lawns.

Momma charges me ten dollars per bag. I don't mind. It's worth it.

Green lawns attract prospective buyers like red carpets attract prosthetic Hollywood celebrities. Here in the suburbs, selling a home is all about curb appeal. Single men, they don't care what lies inside. Appearance, that's what matters. Today, a lawn can make or break a home, leaving a lasting impression regarding how much money—how much green—an owner has stored away in his piggy bank.

Momma doesn't know it yet, but I'm banking on her to find me a man. My green lawn: The perfect pathway to my green heart.

"Sorry, but single men just aren't looking to buy these days," Mr. Real tells me over the phone the next day. "They'd rather not deal with the responsibility. They'd rather rent."

Mr. Real, he gives me the truth. And Momma, I give her cancer. Guess you can say we're into giving and receiving, our social circle.

Me, I receive an epiphany, circling the house for hours after talking to Mr. Real. Me, I tell Momma, once and for all, I'm giving up on giving a poop about finding me some man.

This is until I receive a knock on the door one whole week later. This time Mr. Real brings me someone really nice. No really. "Is this the type of place you're looking for?" Mr. Real asks the nice fellow, leading him into the house.

"Definitely. Just love that lawn," the nice fellow replies.

Me, I'm baking cookies in the kitchen.

Momma, she's baking in the guest room.

Placing globs of cookie dough on a silver tray, I sneak a peek of the nice fellow and gush. He's Green Giant tall with a square jaw, and no circle, no ring, around his finger. "You like cookies?" I ask him, placing the tray in the oven.

"Sure," he responds. "Especially if they're home-made."

"Me too," Mr. Real interrupts.

"Lovely, then you stay here and make sure they don't burn," I inform him.

"But, but . . ." Mr. Real replies. He knows what I'm up to. Deep down, he knows I'm charcoal inside, yearning for a burning, yearning for a chance to be alone with this nice, nice fellow. "Forget about love," Mr. Real calls, as I escort the nice fellow on the remainder of the tour. "Market the house. Market the house!"

The house, the house, I know. But seriously, who gives a hoot about the house? Sure, the wood floors are pretty and made of eco-friendly bamboo. Sure, the afternoon sun shines through the bay window like a god. Sure, Momma and her brown-gravy poop color the lawn green, green, green. But what about my selling points?

I'll be here to greet you whenever you return home from work. I'll prepare dinner every night, I promise. I'll work real hard on making you happy. No need to worry, alcohol will never touch these lips. No sir. With you, I won't need an escape. Plus, I'll never cheat. Never, ever cheat. I won't need to. With you, my heart will remain sugary sweet and fireplace-warm all year long.

"And you can own it all," I tell the nice fellow, leading him onto the final leg of the tour. "My heart, my home, my happiness. All for the bargain price of loving me and only me." And here the nice fellow, here the nice, nice fellow, here's where he looks at me tenderly in the eyes before telling me he thinks he'd rather rent.

ELLIOT

from *Exteriors*

by Billy Masters

Elliot looks younger than his age, 17, and when people spot him out on his own in the middle of nowhere, their first instinct is usually to call the cops, thinking he's some 15-year-old runaway who has stolen his parents' car, which is only about a fourth wrong, he thinks.

He grabs his backpack, pulls out his car keys, and gets in his car, resuming his epic drive across the country from Oregon to Minnesota, to see a guy that contacted him through the Internet through a secret encoded e-mail account that can't be traced. The account user name is earthsaver89, and it's recommended that the user change the password every 48 hours. The 89 is for the year he was born. The earthsaver part is self-explanatory, but this is the first time he has done this, actually contributed to saving the earth.

A girl in Mendocino set it up—this trip that Elliot is on and this guy he's supposed to meet. She lives in a red-wood tree, and that's all she had to tell Elliot to get him on board this strange environmental protection method, which involves actions like blowing up corporate headquarters, torching newly-built housing developments, burning down experiential forests, and driving nails into redwoods to break the lumber company's $100,000 diamond-tipped saw blades.

He thinks that the mysterious group is comprised of the bravest people on earth. He thinks they're probably all pretty hot, too. Especially so because they all have a cause, something Elliot feels that his life needs to be meaningful, to be worth living.

Elliot drives fast down Interstate 80, past the haunting formations of the Badlands. Their bright orange spires and twisted ridges are carved from violent thunderstorms. Their dried mud surfaces remind him of skin, sunburned by hot summer sun. He's sweating and singing along to music as his eyes follow the comical, cartoonish ridgelines of the bright red formations. A pair of Pronghorn, with their vivid black markings, dart through the grass.

The wind is hot and dry, like inside an oven that's been baking all day, and dirty heat from the engine radiates through the vents, making the inside of the car even hotter.

After the Badlands, the land turns flat, and the desert is replaced by flat fields, some green and some brown, already plowed for autumn. He spots a Best Western motel just off the highway exit about ten miles before the Minnesota state line, one of those long narrow buildings with two stories and numbered metal doors in a row. He'd like to check in and turn the air conditioning up to the highest level, but he can't afford a $49.99 room just to be cool. He's used to sleeping in his car to the point that it's almost comfortable, at least when there's a cool breeze or a thunderstorm. It's better than being cold, he thinks.

He parks his car by the stairs and looks around for anyone who might tell him he shouldn't be here. He hates nothing more than being told what to do, unless the person instructing him is someone he wants to be, and he most definitely doesn't want to be a motel manager.

The chemical clean smell of the cheap sheets being washed and dried wafts into the parking lot, making him feel sick and recharging his earlier headache from the tractor exhaust. He thinks about how he's on an environmental mission, about how he's against the horrible stench of cleaning fluids that pollute the ground water. He'll use only clear spring water and vinegar to clean from now on, not that he's ever actually done any cleaning.

In the sizzling parking lot, an old guy with a "Yellowstone Park" hat steals ice from the ice machine to fill the freezer of his RV, its name—Desert Fox—painted on the side of the vehicle as big as a billboard.

A family of hippies, a blond young father, his pretty dark-haired wife, and their two sickeningly angelic little kids—one boy and one girl—load up their VW van, preparing to go either east or west, probably west by the look of them.

Elliot locks his car, grabs his surf shorts, sneaks a perfumed white towel off the maid's cart, and pushes the gate open into the cement pool area. There's no place to change, so he walks back into a vending machine alcove and takes his shirt and pants off. It feels excellent to be naked in this extreme summer heat, he thinks. The slight breeze blowing feels good on his balls. He loves to be naked, and he hasn't felt cool in days, maybe weeks. Plus, he loves to be looked at. He's the opposite of the mousy environmentalists who don't care what they look like and wear dark, drab clothes made of natural materials: brown hemp and tan recycled cotton.

He loves his body; he's even fascinated with it. He works out occasionally, but only because he likes to look good, not for health reasons. His torso is finely muscled and slender in the way that only a teenage boy's body can be. He works hard to look so casually slender, so unnoticeably sexy. This is one of Elliot's contradictions. If asked, he'd swear that he thinks caring about how your body looks is a disgusting act of vanity as narcissistic as a reality show makeover. He knows this part of him is full of shit, but he lives with it.

A 20-year-old girl, all dressed up in a frilly yellow dress for some kind of country formal event, walks into the shady alcove, jiggling quarters in her hand. She stops, shocked, when she sees Elliot, naked.

"Oh, sorry." She instantly turns red and twists around quickly and jerkily, like seeing him naked is wrong.

"No problem." Elliot smiles and steps into his shorts, which are blue and hang low on his hipbones, grabs his dirty clothes in a ball, and walks out into the sun of the pool area. He takes a chaise lounge and leans back, his arms behind his head. He's starting to smell pretty strong. He hasn't had a shower in at least a week. He sits in the sun until a fresh layer of sweat coats his chest, then he stands up and dives into the pool with one clean motion, almost no splash. He's amazed he could be so athletic.

He rubs his hands all over his body, taking a bath in the pool. He looks up and notices a middle-aged woman watching him. He's been caught.

She knows he's just a homeless teenager who hasn't taken a shower in over a week. He has to take off before she calls the cops.

***

Elliot's on the road again, and he's starting to get nervous. At least he's relatively clean now, he thinks, but since he didn't put his pants back on after swimming, his thighs are sticking to the worn leathers seats of the car, and it's an annoying sensation. When he drives over the invisible Minnesota state line on the same monotonous highway, the scenery changes from vast fields of corn and soybeans to little, picturesque patches of still-small pumpkins, artichokes, and watermelons. The air is so humid it's almost raining, even though there are no clouds in the glaring sky.

He takes an exit off the highway and cuts up through some woods and pretty little towns with names like Minnetonka and Winnebago, where the monstrous motor homes are built for selfish old people to drive around spewing pollution into the national parks, caking the stone of Half Dome with creosote.

As Elliot approaches Minneapolis, the traffic gets heavier and the suburbs look the same as anywhere else that's flat. Sitting in traffic in the hot sun annoys him after days of freedom on the road, driving as fast as 90 miles per hour when no one was around. He punches the gas pedal hard, feeling a half put-on guilt about burning fossil fuels, or dead dinosaurs as those in the eco movement say. He thinks about the Twin Cities here and wonders, is one of the cities liberal, one conservative? Is one rich, the other poor? If not, why don't the cities just combine? He can't see a difference from the crowded freeway. Does the speed limit or the regulation that bans skateboarding on city sidewalks change on the city limits?

He exits into downtown Minneapolis along a lake with some suntanned 13-year-olds taking sailing lessons and finds his way to his meeting point, printed out on a green piece of paper from his earthsaver79 e-mail account. The directions are typed out above a map with a red line for him to follow. He follows the red line to a part of town called Uptown, and as soon as he arrives into the shopping area, he can tell this is the area of town where the cool people live.

He finds the exact corner on the map and parks his car along the curb. It's amazing to him that he drove like 3,000 miles to Minnesota, and he found this one street corner in the Midwest.

He turns off his Volvo and thanks it for getting him this far. He hasn't gotten the oil changed in at least 25,000 miles, and he doesn't even know where to add water to the radiator. Engines are a mystery more confusing than life itself. The doors are rusty, and the locks no longer work. The radio's speakers cut out over bumps.

He takes his skateboard out from the back seat and walks to the corner where he's supposed to be waiting for the guy. He sits on his board and watches the city go by.

A boy with a blue Mohawk cruises by on a gas-powered scooter, clutching an iPod and a cell phone in one hand. Two hippie girls skip down the tree-lined street holding hands, smiling and laughing. One has a red bandana in her hair. People doing things, mostly together. Elliot is alone of course.

Elliot picks up his board, jumps onto it, and does some minor tricks on the sidewalk corner, which is connected to the back parking lot of a camping store called Mystery Lake Camping. He flips his board under his feet and turns sharply on two wheels, switching his footing while the board flies just inches above the cement. He skates over to the back corner of the camping store's parking lot and sits on the board in the shade of a huge Maple. He waits for an hour. Maybe this guy isn't going to show. This sucks, he thinks.

He takes a black Sharpie marker out of his pocket and starts to draw a tattoo on his arm—a redwood tree. He gets really into the drawing, so much so that he forgets that he's just waiting for some guy in a parking lot, but the smell of the marker is toxic, bringing his headache on stronger.

He stands up and stretches and throws his skateboard in front of him on the cement.

A guy, maybe 25 years old, with short black hair and wearing a black T-shirt with the sleeves cut off, walks up to him. "Hey. Are you earthsaver79?" The guy's teeth are crooked, and his incisors, or fangs as Elliot calls them, are pointy, making the guy look like a modern-day vampire.

"Yeah. Are you the guy?" Elliot asks.

"Yeah" the guy says, as he points the way for Elliot to walk with him.

Elliot begins to walk, doing as he's told, of course, walking toward his cause.

***

Elliot sits in the guy's room on his bed, waiting for him to get out of the shower or something. He doesn't know where the guy went, but he took his shirt off before he left the room, so he figures probably the shower. He's in an apartment on the top floor of a huge Victorian house, just down the tree-lined street from the camping store.

It's hot and muggy inside the house—no air conditioning and no breeze. The three windows in the bay are closed, their white fabric shades shut, but Elliot can still feel the sun's heat radiating through them.

He looks around.

The room is scattered with shit. Books—*Catcher in the Rye*, some other novels of equal reputation, nothing risky— and magazines—*Outdoor*, *Men's Health* with some guy flexing his abs on the front, *Sierra Club*, *National Geographic* on the floor opened to a story about global warming. The guy's shirt is flung over a chair in front of a messy desk, where a PowerBook's screen saver flashes with scenes of nature—a babbling brook fading into a white sandy beach, then a dirt road cutting through a mossy forest.

Elliot leans down to look under the bed. He sees a metal toolbox full of technical-looking equipment, like wires and little circuit boxes and battery packs, some fuse-type string, a professional-looking stainless steel lighter. He stops looking when he hears a door close somewhere in the hall and falls back on the bed, sick of waiting for this fucking weirdo to come out from wherever he is. He wants to take a shower and sleep on a comfortable bed. Fuck, he thinks, why can't I just be like this dude? I want my own room. I want my own bed, my things scattered around it.

There's a map of the United States tacked to the ceiling. Colored thumbtacks mark certain locations—Vail, Eugene, Marin County.

He looks at big photos tacked on the walls in neat rows—local kids and friends of this guy, he figures. Some girls wearing plastic leis at a party, a teenager wearing a king's crown and no shirt, three guys lined up next to each other on customized, lowered bicycles.

This is the kind of life that Elliot wants but can't have. He's not capable of stillness, of keeping friends long enough to have well-composed photos of them. He can't stay somewhere long enough for the floor to be scattered with familiar and regular things, and this depresses him. He wants to stay here forever; even in this heat, he feels completely at home.

He pulls one of the guy's pillows over his head. It smells like dirty hair, but he likes the smell, and the pillowcase is cool. He takes long slow breaths, trying to calm himself down. He's nervous. He doesn't know what he'll be doing with this guy. Now that he's here, he just wants to fall asleep.

***

Elliot wakes up, freaked out. He's never fallen asleep before without choosing to do so. What is he becoming? he wonders.

The guy is sitting next to him on the bed, fiddling with some wires and needle-nose pliers. He's still a little wet, wearing only a light blue, worn towel. He's wearing a hemp collar necklace thing with a metal pot leaf hanging from it, lying delicately between the muscles of his chest.

Elliot sits up, feels a bit dizzy, and leans forward.

"We have to start getting ready. It'll be dark soon." He glances at Elliot for a millisecond, then his eyes dart back to the wire he's twisting.

"What are we going to be doing?"

"You're doing nothing but looking out for me."

Elliot thinks, I drove this far to be a lookout? His eyes follow a drop of water down the guy's bicep to his elbow.

Without looking up from his little project, the guy asks, "What are you willing to do to save the earth?"

Elliot walks over to the guy's window and looks outside. Two girls are washing their big old American car in the street with a hose. He watches them for a second, then looks back to the guy, who's biting his fingernail, his project thrown on the bed beside him. Elliot starts to feel a tinge of a boner. He's never had sex, but when he masturbates, he almost always thinks of guys, although he's never told anyone this.

The guy watches Elliot closely, like he's waiting for something to happen.

Elliot walks over to the guy and unzips his pants. "I need to know your name." He figures that if he drove this far to be a lookout, he might as well get another item on his life "to-do" list checked off.

The guy opens his mouth. "Fire."

Elliot repeats his name. Fire.

***

Elliot and Fire wear black robes with eyeholes cut out of big floppy hoods. They look like two little boys dressed up for Halloween, and Elliot feels sort of stupid.

It's also fun for him. He can't resist an adventure, and black robes are a good start to any adventure in the most little-boy way. Fuck, Elliot thinks, I was 13 only four years ago. I feel so much older, out on the road on my own, performing an action for the future of the Earth.

Fire carries his red metal toolbox, which Elliot assumes holds electrical timers, fuses, and some explosives, but he hasn't seen them, nor would he know what they look like.

This dude obviously knows what he's doing. He has explosives under his bed. He gets into the driver's side of Elliot's car, so Elliot walks around to the other side. He's never ridden in the passenger seat of his own car, or anyone else's for that matter—he spends too much time alone now.

Fire drives without speaking along the quiet highway—hardly any other cars are out this late in the Midwestern city—then he takes an exit that funnels onto a wide road that winds through rows of brand new houses, all similar, all contemporary takes on traditional houses—watered down Georgians, bland Tudors, vague Italian villas with cathedral ceilings, brass lampposts, curving walkways lined with flowering trees and shrubs in plantings of three. All the windows are dark of each and every house—no exceptions.

Elliot is freaking out a little inside, but he doesn't tell Fire.

Fire just stares straight ahead, finally turning right into a darkened industrial park, small pine trees planted in groves meant to disguise rows of loading docks. He stops the car.

Elliot looks at the clock in the dashboard. It's 3:20 a.m. He steps out. The air has cooled, but not that much. It's still hot, still muggy even though the sun set hours ago.

Fire runs long fuses attached to electrical timers into vents and air-ducts on the perimeter of the chinsy industrial building, which looks like it's made of corrugated steel and aluminum poles and that's about it.

The night is dark and still and incredibly quiet. No animals calling, just the barely perceptible buzz of a power transformer on a pole above him.

Fire runs back over to Elliot and hands him a remote control device of some kind that looks like a garage door opener from the '70s. "Push this button when I tell you to," he whispers, out of breath. He starts to run back toward the building again.

Elliot wants to hug him, talk about nature, ask him about the silver pot leaf hanging around his neck. He wants to drive into the countryside, maybe go skinny-dipping in one the state's 10,000 lakes. He's only been here for about six hours, and he's only driven by one lake. He heard they are warm and there's nothing like swimming in a Minnesota lake in the hottest part of summer.

What Elliot wants and what he thinks he wants are always conflicting, always colliding into a soup that makes him feel unsatisfied, a feeling he's growing used to.

Fire darts across some newly laid sod, over-watered by a sprinkler system that just kicks on as they notice the first black plastic head poking from the artificially green grass, lit by flickering overhead lamps on enormous industrial-strength poles.

Elliot is left standing alone by his car holding the tan plastic box, one big, red, rectangle-shaped button on the upper half.

"Wait," says Elliot. "What are we going to do after?"

"Drive away," Fire whispers and runs toward the building.

Elliot feels used. He thought this mission would feel good, like he was actually changing the world, making it better in some way. Instead, the whole experience feels shallow, just a series of technicalities begun by a weird groping in a stranger's muggy bedroom.

Fire runs back to Elliot from the far side of the office building, smiling. He stops and leans on Elliot's car, out of breath. "It's ready. Let's go." He gets into Elliot's car and closes the door.

Elliot starts the car. "What about pushing the button?"

"Drive."

Elliot starts his car and starts to drive out of the industrial complex the same way he drove in. "Can I stay at your house tonight?"

"No. There's nowhere for you to sleep. I've got room-mates."

Elliot wants to say, "What about your bed?" but he doesn't.

"OK . . ." Fire says when they're about three blocks away from the office building, still in the stupidly land-scaped confines of the office park.

"Push the button."

Elliot pushes the button without hesitating. He's still trying to win over the guy by obeying him, and he hates this about himself. He thinks, I should have just told him to fuck off, told him to push his own little stupid button. Before he even finishes his thought, he hears an explosion.

The trees and the sides of the identical box-like build-ings turn bright yellow, as bright as lightning. A micro-sec-ond later, a loud boom sails through the quiet night, send-ing birds flying from power lines, from dew-wet grass.

Elliot panics. "Well where the fuck am I supposed to go then?" he screams.

"I don't give a shit." The guy's smiling.

Fire started a fire, Elliot thinks.

Fire shoves the remote control into his backpack. "Drop me off here."

Elliot looks to where he's indicating. A bus stop along a suburban street, a couple boring miles from the explosion.

Elliot slows by the curb, and Fire jumps out.

"Peace." He walks away, down a darkened suburban street. Tudors and villas and traditionals, two cars each in every two-car garage.

Elliot feels dirty and raped. He drives out of town before it gets light and sleeps in the back of his Volvo on the side of a lonely country road in the eastern prairies of North Dakota, three hours from Minneapolis. He says "Fire" before he goes to sleep.

CONTRIBUTOR BIOGRAPHIES
(IN ORDER OF APPEARANCE)

MIEK COCCIA lives in Berlin.

BENJI MORRIS lives in Bushwick, Brooklyn in a big old house with eleven anarchists. He's 23 years old, hungry, and has a tattoo of a dog-headed boy on his chest. He wants to turn "Kyler and Wolf-Boy" into a novel by the end of 2008. Contact him at wolf_boy@mac.com.

EDDIE BEVERAGE begs the reader for the correct spelling of "anti-climactic" for his epitaph in the form of a sonnet. Contact him at e_beverage@yahoo.com.

L.A. FIELDS is currently pursuing a degree in English Literature in her home state of Florida, where she is the patron of an unofficial (and highly exclusive) gay independent foreign theater club called Mystery GIFT. Afterwards, she plans to scout out some cardboard real estate and devote herself to writing. Contact her at All4LAF@gmail.com.

BLAIR MASTBAUM is the co-editor of this anthology. He's also going on a week without a shower. He hopes it will rain every day this week. He recently went to Mt. St. Helen's. He lives in Portland, Oregon. He has written and published two novels: *Clay's Way* and *Us Ones In Between*. Contact him at blairmastbaum.com.

PAUL KWIATKOWSKI directs and produces documentaries, films, and music videos. He is currently in the process of completing two collaborative novels due early next year. He apologizes too much and enjoys blacking out,

Ambien, low expectations, gap teeth, heat waves, room service, and films about slutty, complacent teens. Contact him at paul.kwiatkowski@yahoo.com.

JOHN REPOSA was born in 1981 in suburban Massachusetts. Reposa often cites 1984 as the most formative year in his life; in this year he fell down a flight of cement stairs and knocked out several front teeth, which resulted in his having a set of silver caps through much of his early education and his schoolmates calling him names like "pirate face" and "silver tooth." Also, it was in 1984 that Reposa first saw a pornographic movie while attempting to watch a VHS copy of *The Never Ending Story* that had had a portion taped over; he can still recall the scene in great detail and grows wistful at its mention. Reposa now lives in Brooklyn, NY and is currently unemployed and seeking treatment for mild depression and an addiction to pornography. He is working with co-author Paul Kwiatkowski on finishing a pair of novels from which "Muy Simpático" has been excerpted. Contact him at johnreposa@gmail.com.

NICK HUDSON, born in 1981, was raised in a coterie of rural wildernesses across Britain ... settling in Brighthelmstone two years ago, he writes prose, poetry, and music with an autistic prolificacy ... recently signed to Kiddiepunk Records, and having had work published in Dennis Cooper's *Userlands* and Hegarty/Kennedy's *Writing on the Edge*, it appears the catalina has left the bayou for this irrepressible Scorpio urchin. Contact him at bruxo_el@yahoo.co.uk.

BENNETT MADISON is a writer in New York City. He has worked as a receptionist, a phone psychic, and a Markdown Specialist at the Gap, and is the author of several books for young people, including the Lulu Dark mysteries (Penguin/Razorbill) and upcoming *The Blonde of the Joke*, which will be published by HarperCollins in 2008. He attended Sarah Lawrence College, where he majored in craft projects and chit-chat. Contact him at bennettmadison.net.

MICHAEL TYRELL wants you to buy *Broken Land: Poems of Brooklyn* (NYU Press, 2007), an anthology he co-edited with Julia Spicher Kasdorf, so that the book stays in print. He enjoys making abstract-watercolor paintings, visiting obscure parts of New York, reading and writing ghost stories, and watching movies that are painfully funny. He doesn't care for lengthy biographies or people who feel compelled to recite all of their achievements in a single paragraph. According to the porn-name formula (first pet name and first street name), he would be known as Rocky Manhattan. Contact him at michael.tyrell@gmail.com.

SAM J. MILLER is a writer and a community organizer. He lives in the Bronx with his partner of seven years. His work has appeared in numerous 'zines, anthologies, and print and online journals. Visit him at www.samjmiller.com, and/or drop him a line at samjmiller79@yahoo.com.

WILL FABRO is the co-editor of his anthology. He is a graduate of the writing program at University of California, San Diego and has been previously published in *Fresh Men*, selected by Edmund White and edited by Donald Weise, and *Userlands*, edited by Dennis Cooper. His writing has also appeared in *New York Press* and the 'zine *Cheese + Liquor*. He curated and hosted the reading series "This Is Not The New Minstrel Show," which showcased up-and-coming queer writers under 30. Born in 1981 in Hollywood, he currently lives in Brooklyn, New York. He blogs about the New York Mets at smearthequeer.wordpress.com.

MICHAEL GRAVES is most of the following: 1. A literary boy-next-door. 2. A vegetarian. 3. A writer currently crafting a novel. 4. A scaredy cat. 5. A lucky fella who found his partner and best friend, Scott. 6. A 29-year-old. 7. A bunny owner. 8. A daydreamer. 9. A native of Massachusetts. 10. A person who can be found at www.MichaelGraves.blogspot.com.

MICHAEL WOLFE was born in Des Moines, Iowa where he learned to think and hide in his room while listening to The Jets and Billy Ocean cassettes, before his brother turned him on to Quiet Riot and his sister turned him on to Erasure and Yaz. No one knows what he actually accomplished in his room, but his writing should explain some of it. At a certain age, he befriended an invisible boy named Rick and started doing very bad things, like peeing all over the front porch. This is an obvious euphemism that Michael has adopted as a life code. He currently resides in Georgetown, Texas where he maintains one of the most literary and academic bathrooms in North America. It almost doesn't feel like a bathroom at all, but instead a well lived-in library or reading den. There you will find not just one or two of his favorite reads lying on the floor, covered in pubic hair, but also the stale, delightful bouquet of cigarette smoke and cheap aftershave. Visitors often marvel at how un-bathroom-like the bathroom really is. Michael lives with his cat, Boogan, and at this very moment Michael is afraid of him. Boogan is freakishly large and angry. This might explain, in part, why Michael's bathroom looks the way it does. It is the safe haven.

ZACHARY GERMAN is employed by author Tao Lin as a food taster. "Wolf's Teeth" is an excerpt from *Eat When You Feel Sad* (FSG, 2009). Contact him at zacharygerman.com.

WILFRED BRANDT is an American expat living in Sydney, Australia with his awesome boyfriend. He records amazing lo-fi rock under the name Mr. Centipede, writes for some magazines, works at a library, swims in a Speedo, and still skateboards (occasionally). He's working on his debut novel entitled *Cannonball* and hopes to one day own a dog. Contact him at wilfredbrandt@juno.com.

STEPHEN BOYER: It's late afternoon and I am naked on a beach with a bottle of white wine and a bowl of spinach, red onions, beets, and croutons covered in balsamic vinaigrette dressing. A few people are around. They join me, and we build a fire and talk late into the evening. Contact him at stephenjboyer@gmail.com.

ANTHONY PAULL is a freelance writer and syndicated columnist from Sarasota, Florida. His column "The Dating Diet" is published in several GLBT newspapers across the United States. His first short story was published in *Best Date Ever: True Stories That Celebrate Gay Relationships* (Alyson Books). He has just completed his first novel. Contact him at anthonypaull.com.

BILLY MASTERS spent his late teenage years living on the streets by choice. He now resides in a small apartment owned by a hippie collective in Arcata, California and participates in covert environmental actions. He also writes, takes photographs, and works out. Contact him at hipsterhater@mac.com.

ACKNOWLEDGEMENTS

BLAIR MASTBAUM thanks Don Weise for conceiving this book and giving him the chance to co-edit it. Thank you, Greg Jones and Jon Anderson at Running Press. Thank you, rainy days of Portland, and thank you, Scott Coffey, for reading this for me and helping me with selections.

WILL FABRO thanks Dennis Cooper, Corrine Fitzpatrick, Eileen Myles, Laurie Weeks, Don Weise, and Edmund White.

WEBSITE

Visit the Cool Thing blog at coolthingbestgayfiction.blogspot.com.